THE REBUILDING OF A
LOTUS ELAN

SUPPLEMENT ENGINE SECTION

BRIAN BUCKLAND

ISBN 978-3-96473-004-6 (binded Supplement Engine Section)

Published by
Grafos Verlag GmbH
Frauenrichterstr. 2
92637 Weiden
GERMANY
Phone +49 (0) 961 / 20490400
Email info@grafos-verlag.de

Printed by LIGHTNING SOURCE LLC.

SECTION E.

THE ENGINE.

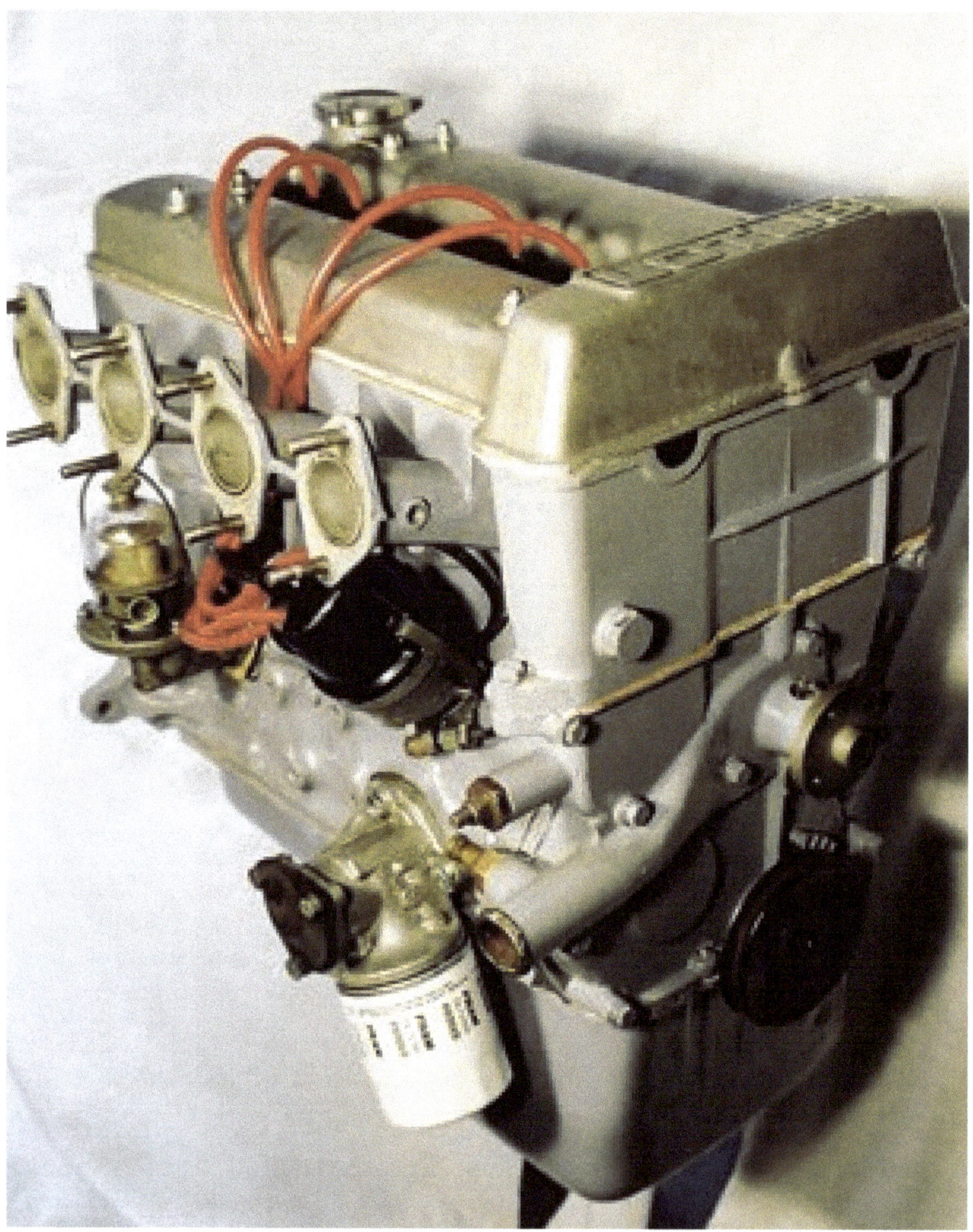

LOTUS TWIN CAM ENGINE.

A twin cam engine just rebuilt and ready to install in an Elan. 1558 cc.'s of motoring pleasure about to be unleashed. They are tight to fit and some degree of patience is required with a bit of gentle easing. It does help if you have done it a few times before. This is a Mark 2 engine with a 6 bolt crankshaft for the flywheel. It is a Special Equipment version and should have a green hammertone finish rocker cover.

GENERAL.

The Elan engine is a superb power unit. It is basically a Ford Cortina 1500 block to which Lotus added a twin overhead cam aluminium cylinder head, of their own design, dramatically improving the breathing and therefore power output of the engine. The new cams were driven by a timing chain enclosed in a new timing chest on the front of the engine with a timing chain tensioner. This timing chest also held the water pump which became the Achilles heel of the engine. The valves at an angle of 27 degrees sat in a hemispherical combustion chamber and the large induction and exhaust tracts were straight in and out keeping the resistance to gas flow to a minimum. Tappets were shims in buckets (followers). The old push rod cam shaft was retained as a jack shaft to drive the petrol pump, oil pump and the distributor.

The first 21 production engines were 1498cc and Lotus quoted the output as 95BHP. The first cars were therefore, not surprisingly, called a Lotus Elan 1500. The size of the engine was increased to 1558cc with a power output of 105BHP. This engine size was decided upon for racing purposes for the 1600cc saloon class which the Lotus Cortina raced in. There was a racing rule allowance to rebore the standard block 1mm (40thou) oversize and this brings the 1558cc twin cam engine to 1598cc, coincidentally just right for the class. Chapman playing politics helping Ford.

To help sales along Lotus produced a Special Equipment Elan and the SE engine had up-rated cams and modified Weber carburettors. The output of this engine was 115BHP.

In a final effort to increase flagging sales after 8 good years the S4 car was given a real performance workover for the Elan Sprint with a "Big Valve" engine producing 126BHP. As the name suggests the inlet valves were larger diameter, the cams were up-rated, compression ratio was increased from 9.5 to 10.5 and the carburettors were re jetted.

There were various small modifications to the engine for different markets, the largest being the exhaust emission engine for the American market to meet the new Federal Regulations from 1968. To economise Lotus produced a new 2 port head with Stromberg carburettors. There was a market reaction against this and Weber carburettors were reintroduced.

Leaving out the cylinder head and its timing chest components, Lotus produced their own crankshaft and pistons for the engine but all the other ancillaries were Ford based for mass produced economy. The engine was over-square with a stroke of 2.86 inches and a standard bore diameter of 3.25 inches allowing the engine to rev. pretty freely.

In the 1960's, if you had the Elan throttle down, you did not have to look in the rear view mirror, there was nobody behind. 50 years later with modern 4 valves a cylinder engines, modern cars overtake with normality especially on motorways (but not on a winding road) However, the Lotus twin cam engine will go down in history as a tremendous power plant leading the way for other manufacturers. It was in its time the only 4 cylinder twin overhead cam engine in production in the UK.

The engine for some reason has a reputation for unreliability, which is a nonsense. Yes there are some problems, which are dealt with, but the biggest problem the author has found are owners tinkering with the engine whilst not being sure what they are doing and then blaming the engine when it goes wrong. It is a fussy engine and everything has to be right for the engine to perform as it should.

Despite the various engine formulations above, the basic engine is assembled the same way and this procedure is described in detail. The normal reasons for engine restoration work on the 'twink' as any engine being:
- General wear giving low oil pressure together with burning oil.
- Valve and valve seat wear giving uneven firing.
- Timing chain adjustment problems
- Water pump failure.

Low oil pressure is normally caused by oil escaping passed worn bearings. It can be the oil pump and this is worth checking by substitution before the expense of full engine work. If the problem is bearing wear a new oil pump with the now required new bearings and machining is a good idea anyway so this trial has not cost you additional money. Do not rely on the dash mounted oil pressure gauge for this test as these gauges are notoriously inaccurate. I have a 4 inch diameter accurate gauge for problem oil pressure checking.

Oil burning, characterised by blue smoke coming from the exhaust can either be worn piston rings and cylinder bores and/or worn valve guides. An engine leakage test will confirm how well or otherwise the piston rings are sealing in the bore.

Valve problems can be checked by a simple compression test. Compression can be affected by ring and bore wear but not as much as leaking valves. Valve clearances can also affect compression if the clearances are not checked and the tappets re-shimmed every 6000 miles or so as recommended by Lotus. Seat and valve wear will reduce clearances until the shim can and will hold the valve open. The design of the valve gear is good and even in these days of unleaded petrol and increased valve seat recession the 6000 mile check does not need to be increased.

The timing chain is a hard working component and the resulting wear can be taken out by adjustment of the chain tension. This is a simple job to do. When the adjustment is nearly fully screwed in then a new chain is required. Sprocket wear also affects chain tension but chain replacement is much more frequent.

The replacement of a worn water pump while the engine is in service can be a nightmare for the amateur owner. This procedure is described later in detail. You need a hydraulic press to insert a new pump assembly into the timing chest and whilst a dealer will do that for you this is only a small part of the total job. I do not exaggerate when I say this has to be the worst job you can do on an Elan. It is always a good idea to replace a water pump during a strip down to prevent a possible replacement in service.

ENGINE IDENTIFICATION.

Ford Crankcase.
All blocks destined for the Lotus twin cam were tested by Ford for their cylinder wall thickness and stamped on the front right hand face LAA, LA or LB. LAA was the thickest and the 'L' we can assume for Lotus.

To help Ford to differentiate between their own and blocks destined for Lotus, Mark 2 engines had a large letter L cast between the fixing bolts of the engine mounting boss on the right hand side

The main Ford block identification number is on the left hand side of the engine, just behind the mounting bracket bosses and about 1 inch above the flange of the sump connection.

Early Mark 1 Engines: Ford 116E, up to Lotus No. LP 7463.
The 116E was ground off by Lotus leaving **6015**

Late Mark 1 Engines: Ford 120E, up to Lotus No. LP 7799.
The 120E was ground off by Lotus leaving **6015.**

Early Mark 2 Engines: Ford 120E, from Lotus No. 7800
The 120E was ground off and over stamped 3020 giving **3020 6015**

Later Mark 2 Engines:
the Ford identification was left on the block, which was **681F 6015.**

Final Mark 2 Engines:
the Ford identification was left on the block, which was **701M 6015.**

Lotus Engine Numbers
The Lotus engine number was stamped on top of the right hand engine mounting bracket boss. All twin cam engine numbers up to 1968 are identified as LP**** (Lotus Part) The engine number was repeated on the rear face of the cylinder head on the small upstand around the core plug. These days when 'originality' is important these stamped engine numbers, obviously, should be the same. This number is difficult to see with the engine installed with the cylinder head rear face within 1 inch of the engine bay/cockpit bulkhead.

During 1968, coinciding with the launch of the Plus 2 and the factory move to Hethel, Lotus were becoming a much more mature manufacturer and began to identify engines more closely with a letter prefix instead of the LP designation. According to Miles Wilkins in his Twin Cam Engine book Lotus made 24 variations.

Engines had a further identification letter stamped on the round boss in the plug well just forward of No. 1 cylinder from 1968 onwards, a H, N or S.
H, High Compression.
N, High compression and Big Valves (Sprint Cars)
S, Big Valves standard compression

ENGINE IDENTIFICATION (1)

The stamping on the top left front face of the block LAA, LA or LB denoted the thickness of the casting. The thickness of the metal dictated how much it could be machined out. LAA was the thickest (and most sought after) going down to LB the thinest.

The 'L' between the engine mounting shows that although cast by Ford it was made primarily for Lotus and the twin cam engine.

ENGINE IDENTIFICATION (2)

Early Mark 2 Engine.
The Ford 120E was ground off and over stamped 3020. This block was not done very well but you can just see it.

Later Mark 2 Engines had the Ford identification Marks left on.

The final twin cam blocks also had the Ford identification numbers left on.

Note the flywheel mounting boss. It has a dowel which is not usual with a 6 bolt crankshaft. There is a drilled hole for a dowel but it is not fitted.

LOTUS ENGINE IDENTIFICATION NUMBERS

These are stamped on the RHS of the engine support boss. You can see in the photograph the 4 No. Fixing holes of the resilient mounting.

Without the L identification between the 4 fixings this is probably a 1966 4 bolt crankshaft engine from a very early S3.

CYLINDER HEAD

A photograph of the rear face of an early head showing the engine number stamped on the boss of the core plug. This number should match the number stamped on the crankcase above the RHS engine mounting. This head is probably from a Series One Elan

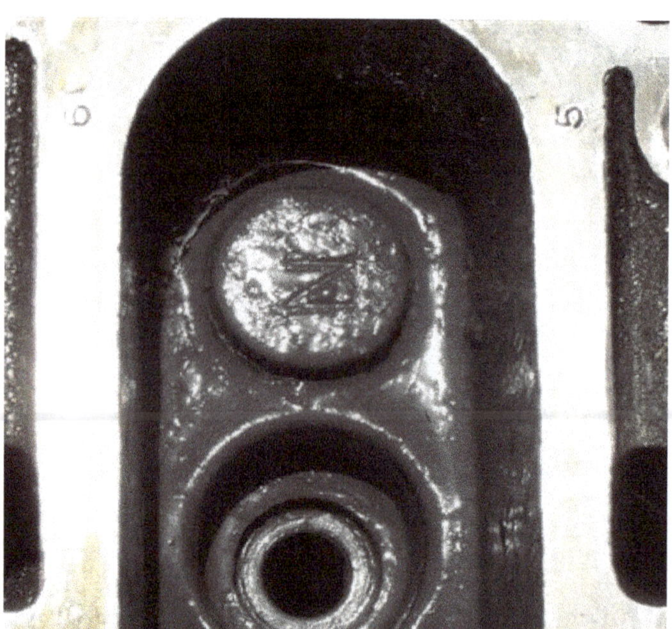

ADDITIONAL IDENTIFICATION

From 1968 additional identification letters were stamped on the round boss in the head plug well. Either a 'H', 'N' or 'S'. Please see text. This head stamped with a letter N indicates a Sprint head.

BRIAN BUCKLAND - THE REBUILDING OF A LOTUS ELAN

CAM COVERS FOR ENGINE IDENTIFICATION.

STANDARD.

As fitted to the first Elans, an original hammertone blue cam cover for the standard Weber head, B type cams giving 105 BHP.

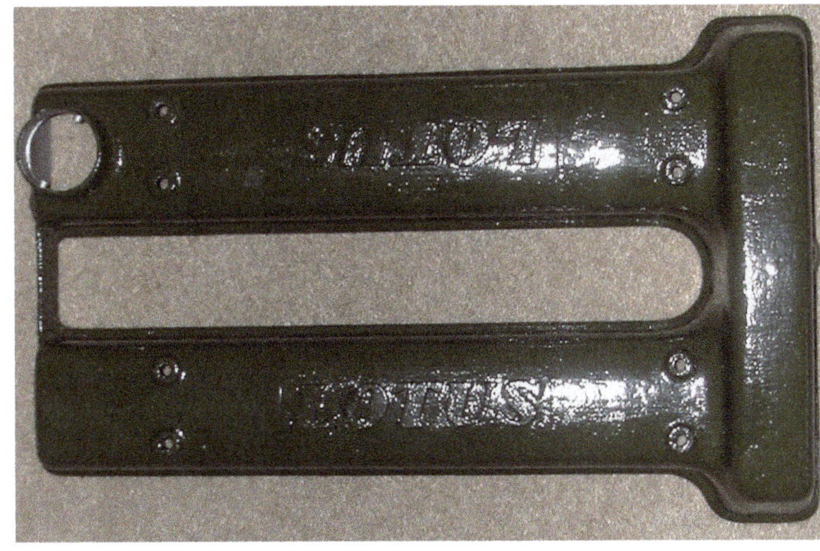

SPECIAL EQUIPMENT 1965

Should be Hammertone but the colour BMC green is correct for the Special Equipment Elan twin cam. This cam cover will look better when the *LOTUS* identification lettering is polished.

115 BHP, C type 2 ring cams and Weber carburettors.

STANDARD STROMBERG CARBURETTOR CAM COVER, 1969.

Photographed at a show and on the wrong cylinder head. C type 1 ring cam. Crackle black finish which should be matt black. It looks very nice though.

CAM COVERS FOR ENGINE IDENTIFICATION (cont.)

RED CRACKLE CAM COVER.
Several Elan twin cams had this cam cover.

1968 Super SE with Weber carburettors and D type cams.
1969 Federal Elan with Stromberg carburettors and C type cams.
1970 Federal Big valve Elan, Stromberg carburettors and C type cams.

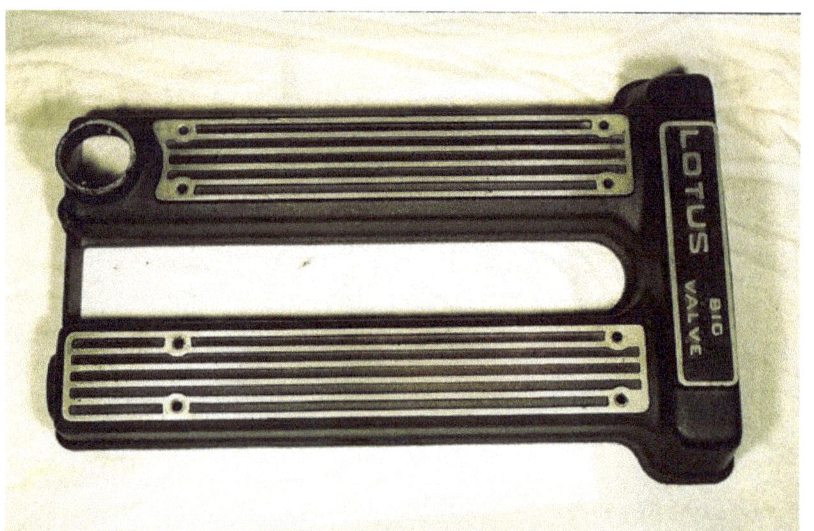

SPRINT FROM 1971.

The Elan Big Valve engine crackle black with flutes.

Weber carburettors with D type 2 ring cams.

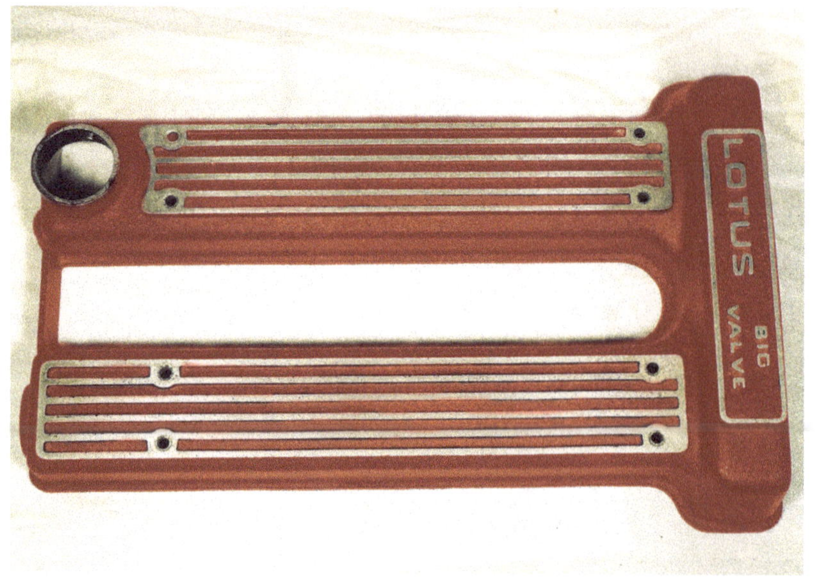

SPRINT FROM 1972. (European model).

The Elan Big Valve engine crackle red with flutes.

Dellorto carburettors with D type 2 ring cams.

If in the event of connecting rods wanting to see the light of day and Lotus supplied a replacement crankcase as a spare part, no Lotus engine number would have been stamped on this block.

CAM COVERS.

Lotus produced 5 No. cam covers for the twin cam engine;

Original cover with *LOTUS* in italic print over the cams.
A cover with LOTUS across the top of the raised timing chain section.
A cover with LOTUS Big Valve across the top of the raised timing chain section and polished flutes over the cams.
A cover for the Stromberg carburettor engine which had a 2 bolt boss over the middle of the inlet cam for the throttle return spring bracket.
A cover for the Europa with a drive for the generator pulley extending from the rear of a cam shaft.

The early cam cover had a round oil filler cap which was changed for later Series 3 cars to a 3 eared type for a better grip due to the difficulty of removing the round type especially when hot.

We are only concerned with the first four cam covers which were fitted to Elans. In addition to the aluminium casting differences, they were finished in different colours denoting the type and power output of the engine. See photographs in this Section.

CAMS.

There were 4 No. types of cam fitted to the Elan twin cam engine, basically increasing power output as the engine was developed over the years. These cams can be identified by a machined groove on the sprocket connection boss.

Engine / car model.	Cam.
Standard Elan.	B Type, no grooves.
Special Equipment.	C Type, 1 groove.
Super Special Equipment/ Big Valve.	D Type, 2 grooves.
Federal Big Valve.	E Type, no grooves.

See photographs in this Section.
Cam lift is the difference between the cam base diameter and the measurement from the base heel to the lobe tip less also the tappet clearance.

CAMSHAFTS

Standard twin cam.

B Type.

No rings on sprocket boss.

Special Equipment.

C Type.

Single ring identification.

Big Valve / Sprint.

D Type.

2 ring identification.

NORMAL PROBLEMS.

Water Pump and Timing Chest
The water pump has an awesome reputation for being short lived and for the unbelievable difficulty in replacement. On average mine last 40 - 60 000 miles, 2 to 3 years for me, and I will probably replace twice in between full engine rebuilds. The tension of the fan belt drive is critical and should not be tighter than the ½ inch movement specified by Lotus. If anything, run it a shade slacker.

A modification to the timing chest is available today which allows the pump to be withdrawn from the chest in a module for easy replacement. This is a super idea but is not cheap. However, if you have to replace a broken timing chest anyway it is probably the best way to go, especially so if you use your Elan frequently rather than 100 miles annually during the summer holiday! I have had a timing chest with a module water pump in my normal road going Elan now for 2 years and am very pleased with it, just coming up to 30,000 miles.

Problems with the timing chest include;
Broken off generator/alternator belt tension bracket boss
Broken or more likely corroded away housing for the pump seal
Split in casting around the oil level pipe. (Due to force fitting normally when cold)
Stripped sump threads but these can be repaired.
Warped out of shape due normally to ham fisted removal from the block.

Fan belt generator adjustment bracket.
The early timing chest boss for the slotted adjustment bracket of the fan belt is a little fragile and can crack and break off. The later timing chests had a much stronger boss. Many of these early broken timing chests are seen for sale at Part Fairs and are usually ignored by customers. However, they can be well repaired by cutting/filing off the remains of the boss down to a flat and making up a steel replacement which can be bolted through the wall of the timing chest to the timing chain damper steel backing. Do not drill the bracket hole until the new boss is bolted to the timing chest and can be aligned with the hole in the timing chest backing plate.

Alternator
The original generator produced 22 amps maximum and in the middle of winter this could be a fraction short of that required. I recommend fitting a 35 amp alternator which solves all the starting problems on a cold morning. When fitting alternators to the twin cam, Lotus further strengthened the drive belt adjustment engine boss with a bracing steel strap across it. The belt tension adjustment bracket was U shaped so it connected to both sides of the timing chest to give a much better support against the belt tension. See SECTION M, Electrical Equipment.

Some owners fit much larger alternators in the mistaken belief that this is better. The car only needs a 35 Amp size. You must also understand that if the alternator produces 60 amps for instance, the cabling must be of a size to accommodate that load.

Cam followers.
The early engines had the cam followers (buckets) running directly in the aluminium head and this caused wear problems. Large clearances caused by the wear allowed

excess oil to reach the valve stem. The valve guide was short by conventional standards and if worn, would allow large quantities of this oil to be burnt in the combustion chamber. The absence of valve stem seals did not help. Lotus inserted steel liners into the head to cure the problem, the hotter exhaust side first followed reasonably quickly by the inlets.

Oil leaks.
Oil leaks are another problem and the worst of these is the timing chest joint to the back plate. Maybe Lotus can tell me why the original Ford timing case and water pump had a gasket but the larger twin cam component did not. A metal to metal joint? This problem is helped along nicely with higher than normal crankcase pressure. Even newly machined engines seem to have a goodly amount of pressure and this can blow the oil asunder.

The early Mark 1 engines, installed in the S1, S2 and half of the S3's, which had a 4 bolt flywheel, had a rope main oil seal on the crankshaft and these could be tricky. Ford must have had production problems also, because around 1967 they changed to a rubber lip seal held in a modified housing, exactly as the front end seal but larger. This totally cured the problem at a stroke and these later engines with a 6 bolt crankshaft became known as the Mark 2 unit.

The cork saddle gaskets can be difficult especially at the front end where the sump bolts screw into the Lotus timing chest. These threads have a strong tendency to strip leaving the joint on the slack side. The now universal use of coil insert thread replacements have certainly helped to cure this problem, the stainless steel insert being much stronger than the original aluminium.

Connecting rods.
Early connecting rods, the L type Cortina 116E with 11/32 diameter bolts had a reputation for wanting to see the light of day through the crankcase. To be fair this was normally when revving the engine above 6000 rpm. Lotus changed them during later production of the Mark 1 engine to C type 125E with 3/8 diameter bolts. It is a very good idea if you have bought an early car with a Mark 1 engine and L Type 116E conrods to change them to the 125E rods which are readily available.

Engine mounting chassis brackets and resilient mountings
There are 2 problems with engine mounting brackets, the first is really a chassis problem but it is a good idea when working on the engine to keep a good look out for cracking across the top of the Lotus chassis bracketry. This crack is very common and does need to be reinforced and welded up.

The second problem is another that Lotus should have sorted with their chassis. On Weber and Dellorto carburettor cars, the right hand side resilient engine mounting bracket is not the same as the left. It is higher mounted lifting the engine on this right hand side so the unit is slightly canted over. It is a specially made mounting therefore not as economic as the mass produced standard Ford bracket fitted on the left hand side. If the proper bracket is not fitted the carburettor air box rests on top of the footwell bulkhead causing very unpleasant vibration and eventually damaging/cutting a hole in the glassfibre.

ORIGINAL CONROD (Below).

The original twin cam Ford 116E conrod with 11/32 bolts which had the reputation for easing out through the side of the crankcase at anything above 6000RPM. Lotus changed them to 125E rods with the larger 3/8 big end bolts. If you have bought an early car which still has the 116E rods you would be well advised to replace.

ENGINE MOUNTING STAND (Below).
This is an engine mounting stand I made up for myself. Proprietary stands (normally universal) are made to support the engine from the rear where the bell housing bolts go. This prevents any work being carried out on the flywheel / clutch assembly whilst on the stand. I saw a photograph of Lotus building engines with the engine supported from the left hand mounting. I thought that this was better and built mine likewise. It has supported a few 'twinks' in its time.

The Stromberg carburettor cars had the standard Ford bracket fitted both sides.

On the Spyder chassis there are extra brackets that bolt directly to the engine to get over the bell housing 'jam' problem. These brackets also allow for the difference in height required and the cheaper standard resilient mounting can be used each side.

What is not so well known is that the closed end on the resilient mounting also acts as a safety device, and prevents the engine falling out of the engine bay in the event of the rubber mounting breaking its bonding. This is the reason why the mounting is one way up bolted to the engine for the Lotus chassis and the other way when bolted directly to the Spyder chassis instead.

The rubbers do soften on these resilient mounts, oil seems to do them no good at all, and it is a very good idea to keep a watchful eye on them.

BUYING AN ENGINE.
You cannot go to a Lotus show and not see engines for sale. The author has bought some engines from this source. If you buy a wrecked car for restoration you are in the same problems with the engine, you just do not know what state the engine is in or what components are installed on the inside.

To add to the confusion all cam covers will fit other twin cam engines. The colour of the covers is one of the main external identification features of the engine. All camshafts will fit other engines. The higher lift of the Big Valve camshaft is where the majority of the extra power comes from. A Big Valve head will fit a standard engine block and a standard head will fit on a Big Valve engine. A 4 bolt crankshaft is interchangeable with a 6 bolt component if you change the sump and rear seal housing. If you are buying a specific period engine for a specific model Elan, yes, take note of the crankcase identification numbers, but the only way to be totally sure of what is in the engine and its condition is to strip it.

Some sellers will know exactly what they are selling, the majority will not. Some are honest some are not. Engines involved in engine bay fires are to be totally avoided. The aluminium head is odds on to be ruined, the aluminium will be softened and NO AMOUNT of time and trouble will rescue it. On the plus side, with the machining techniques and skills available today, most wrecked engines can be machined and rebuilt to as good as, or if not better, than the original standard. Even cracked heads can be aluminium welded and re-machined.

TORQUE WRENCHES.
The engine requires bolt tightness from 6 lb.ft to 65 lb.ft. This is too much range for one torque wrench. To use a wrench right on the end of the scale, especially on the small settings, is not good engineering. All torque wrenches will have a percentage error, normally in the region of +/- 3%. This error will increase at the extremities of the range. I have 3 No. wrenches, a small one with 3/8 inch drive that I use for the ¼ and 5/16 bolts, a second one which will cope with head bolts and a third one used for wheel hubs. When you have finished with a torque wrench, always wind it back to the minimum setting to avoid any fatigue of the spring.

ENGINE MOUNTING BRACKETS.

Resilient Ford Mounting (left).

Lotus used with their chassis a standard Ford resilient mounting on the left hand side of the engine and a modified mounting to give a higher support on the right hand side. The closed bracket at the top was a safety measure to stop the engine falling out in the case of bonding failure. The mountings bolted directly to the engine with 4 No. 5/16 UNC bolts. The left hand side also had 1.125 inch long spacers.

Many Lotus owners have fitted nice new standard brackets both sides and puzzled why the carburettors are vibrating on the footwell.

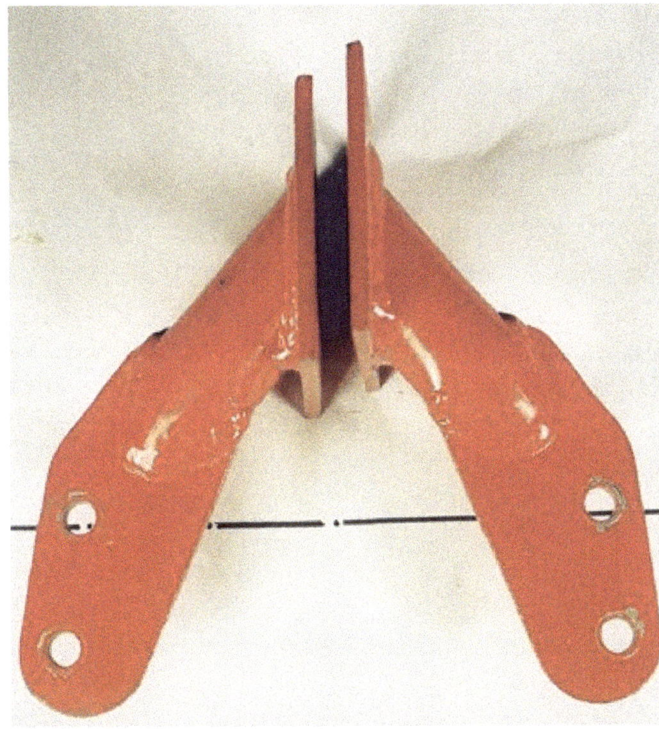

Spyder Chassis (left).

To get over the engine height problem and also the bell housing 'jam' problem Spyder install some extra brackets. The STANDARD Ford resilient mountings are in this case bolted to the chassis upside down, See also Section A, The Chassis. The extra brackets which adjust the engine height are bolted direct to the engine. In the photograph I have drawn a line on the white backing sheet in line with the top resilient mounting bracket holes. You can clearly see the difference in height of the engine mounting plates. The 1.125 inch spacers are still required on the left hand side.

BOOKS TO INWARDLY DIGEST.

The Lotus workshop manual goes through the rebuilding of the twin cam comprehensively. Full torque requirements are scheduled and as already noted the engine rebuilder is well advised to take note of these.

The Miles Wilkins book, The Lotus Twin Cam Engine is a most interesting story of the engine together with instructions on how to put the engine together and practical tips to make those tricky jobs seem easy. However there are gaps. I would not contemplate the rebuilding of a 'twink' without an engine stand.

REMOVING THE ENGINE.

Removing the engine is straight forward and if no problem is incurred with rusted nuts and bolts should take about 3 hours for the average home mechanic. The engine can come out complete with gearbox or on its own. Normally I remove complete with the gearbox as I think it is easier to re-install that way. There is, as with any engine removal, some preparatory work.

1. Disconnect the battery. Always the first job.
2. Remove the bonnet to a safe storage area.
3. Remove the inlet manifold vacuum take off to the headlight reservoir T connection.
4. Remove the vacuum pipe connection to the brake servo if fitted.
5. Drain the radiator, remove hoses including heater hoses and disconnect electrical supplies if there is an electrical radiator cooling fan.
6. Remove the radiator.
7. Disconnect the heater control cable from valve.
8. Remove the choke, remove the throttle.
9. Remove the carburettor air box cover.
10. Disconnect the flexible petrol line from pump to carburettors.
11. Remove the carburettor metal back plate.
12. Remove the carburettors. Blank the induction ports with clean rags. I use bubble wrap these days.
13. Disconnect the petrol feed to the petrol pump.
14. Remove the electrical connections to the distributor including HV. and LV. cables.
15. Remove the distributor and blank hole with clean rag.
16. Disconnect the starter motor cable.
17. Remove the starter motor.
18. Disconnect the reversing light switch electrical connections if fitted, S4 onwards.
19. Disconnect the speedometer cable from instrument, pull through bulkhead and tape to the induction manifold out of the way.
20. Disconnect the oil pressure line from the engine adjacent to the distributor position. Coil up and tape to prevent damage.
21. Disconnect the generator/alternator cables, remove generator/alternator and fan belt.
22. Remove the anti roll bar.
23. Remove the chassis cross brace. The early Lotus chassis cross brace is not removable but should be ground off and replaced by a bolt on one (see Chassis Section)

24. Remove the exhaust system. If there is a fabricated exhaust manifold, this will not come out whilst the engine mounting bracket is attached.
25. Put a piece of wood under the engine sump to spread load and support engine with trolley jack. Remove the left hand resilient mounting in a stressed skin chassis or the mounting bracket in a Spyder space frame chassis.
26. Remove the engine exhaust manifold and replace the engine mounting leaving the nuts slack. Blank the exhaust ports with clean rags.
27. Loosen the clutch slave cylinder bleed nipple, fit a bleed pipe and pump the fluid out into a container, gently.
28. Disconnect the clutch pipe. Remove slave cylinder. Blank clutch master cylinder with 3/8 UNF bolt to prevent dirt ingress.
29. Drain the engine and gearbox oil.
30. In the car, remove bottom dashboard switches. Not totally necessary but it does prevent damage to them.
31. Remove the gear lever knob and rubber boot.
32. Remove the chassis spine console.
33. Unscrew the plastic gear lever retainer and remove lever. Blank gearbox hole with a clean rag.
34. Support the gearbox with the jack and remove the gearbox rear support.
35. Remove the engine earth strap. It may already be off as it usually attaches to the engine mounting or the generator bracket.

The engine is now ready for removal from the car.

Lifting Out the Engine

Do not lift the engine by the induction manifolds. Whatever your engine lifting arrangement is, put a wire or strong rope strop under the engine on the rear higher part of the sump and connect to the lifting eye. Put the strop through the induction manifold on the right side of the engine and the extended engine mounting on the other (the one you have just replaced) This ensures that nothing slips. The engine comes out at approx. 30 degrees angle so the strop wants to be as short as possible because of the height you have to lift the engine for the gearbox tail to clear the car.

Take the weight of the engine with your lifting gear. Remove the engine mounting bolts. Lift the engine slowly. The front of the sump is very close to the front cross member and as the sump just clears this, pull your lifting gear forward, or push the car backward. The next thing to happen as you continue to slowly lift and ease the engine forward is that the gearbox tail clears the chassis spine and can bang down on the ground if you are not careful. I always put a piece of 3 inch foam under this area to make sure there is no damage. You probably will get the remains of the oil in the gearbox come out of the propeller shaft connection. It can be messy but it is very difficult to catch in a container under a very low car.

At exactly this time the bell housing has to clear the chassis engine mounting brackets. You should have enough room now to lift the engine for the bell housing to clear. This bit is known as the 'bell house jam' The beast is nearly out now, ease the engine a bit further forward then a big lift until the gearbox tail is clear of the body. Roll the car away. Lower the engine onto a flat surface and unbolt and remove the gearbox. If you have an engine mounting stand lift the engine onto it and remove the lifting gear.

INSTALLING THE ENGINE.

It is tempting to say that installing the engine is done in the reverse way to removing but there is a little more to it than that, so first the preparation. I always install the engine and gearbox as a complete assembly. The engine is (or should be) rebuilt and gleaming, certainly clean as if a repair has just been done. Cover all holes by wedging clean rags in them, we do not want any nonsense at this stage of the game. Fit the speedometer drive cable and tape out of the way onto the inlet manifold. Do not put any oil in the gearbox or it will all drain out when the engine is suspended at an angle. Install the propeller shaft in the chassis spine but do not connect to the differential, slide it up and alongside the differential so as not to interfere with the gearbox as it enters the chassis spine. Please note that with a Spyder space frame chassis the prop. shaft can be installed after the engine. Do not connect the following to the engine/gearbox; gearbox chassis support, the gear lever, carburettors, starter motor, generator/alternator, any exhaust manifold or clutch slave cylinder.

Support the engine in a strop so when lifted it hangs down at about 30/35 degrees towards the gearbox. (See photograph)

With the engine over the engine bay begin to lower the engine, guiding the tail of the gearbox towards the recess under the scuttle towards the spine of the chassis.

At this stage when the gearbox tail is adjacent to the scuttle recess the engine can start to be eased rearwards, lowering as well so the tail points towards the spine chassis entrance.

We now come to the infamous 'bell house jam'. On a Lotus chassis the bell housing is too wide to go through the chassis engine brackets and must go over the top of these brackets. The engine has to be eased rearwards before final lowering so that the maximum width of the housing JUST clears (I will say that again JUST clears) the engine mounting brackets. At EXACTLY this time the gearbox tail will be JUST clear of the spine entrance but due to the angle the engine is being supported at, it will be lower. Put a trolley jack under the gearbox and lift, this should bring the gearbox tail to point straight down the spine tunnel. (See photograph)

Ease the engine rearwards into the spine lowering at the same time. The gearbox should roll on the trolley jack.

If the prop. shaft is fitted, the prop. shaft yoke has to be fed into the gearbox as the engine is eased rearwards. Remove the right hand side seat for this operation and peel down the centre section of carpet which should not be stuck to the GRP. You will now have access to the spine opening which is closed with a large rubber grommet. Support the yoke and as the engine is eased rearwards slide the yoke into the gearbox. It is wise to grease the seal running surface to ensure the first run is not dry.

Connect the engine mounting brackets. Remove the lifting strop.

Connect the gearbox resilient mounting.

Connect the gearbox support to this mounting do not tighten the nuts.

Lift the gearbox with a jack and connect the gearbox support to the chassis with the 4 No. bolts.

Tighten the resilient mounting to the support.

The engine is in and the rest is just fitting the ancillary items. It is necessary to remove the left hand resilient engine mount to install the exhaust manifold. Jack the engine from the sump. Use a stout piece of wood between sump and jack to spread load and prevent damage.

NOTE: Connecting the propeller shaft to the gearbox.
If the engine has been removed for a running maintenance job, such as replacement water pump described following, it is likely that the propeller shaft has not been disconnected from the differential. The yoke can be fed into the gearbox as described above. It is a 2 man job, one man supporting the yoke in the right position from the chassis access hole and the other easing the engine rearwards and 'wobbbling' it at the same time to line up the splines. Always grease the yoke shaft so that the first run is not dry.

If the prop. shaft has been removed it is easier to reconnect after the engine has been installed. On a Lotus stressed skin chassis lay the shaft in the spine but slide the rear up and over the differential. When the engine is installed ease the shaft forward and into the gearbox. Connect to the differential. If the differential is not fitted the shaft can be installed from the rear when the engine is in place.

On a space frame chassis, the prop. shaft can be installed after the engine has been installed. The shaft slides into the spine through a specially designed access slot in the bottom of the chassis before the exhaust is fitted.

Sometimes the access hole in the GRP does not line up with the holes in the chassis making access difficult. It would not be the first time that I have had to correct this. A worthwhile job to do.

Angle of Dangle (Above)

Note the angle the engine goes into the engine bay. Lotus say 30 degrees. For the first time I measured the angle with an adjustable spirit level. This one was 35 degrees and slipped in without a problem. Note the neoprene sheet protecting the GRP and paintwork. Note also the rope strop, it is tied at the top with a sheet bend for any seamen who also build Elans. Please make sure the strop is secure.

Entry of Gearbox into chassis spine (below)

A small trolley jack lifts the gearbox so that the tail can enter the chassis spine. When the engine is pushed backwards the whole assembly rolls on the trolley jack. Easy.

REBUILDING THE ENGINE GENERAL

Rebuilding steps for the engine are clearly numbered. Explanations of tricky tasks are also given to help the "first timers". The photographs are also designed to give visual aid and to remove doubt. The rebuild is described in four parts, the cylinder head, the timing chest/water pump, the major assembly of the engine and then a fully illustrated guide of each assembly step. There is also further essential help including **Starting the engine for the first time** and **Replacing a water pump with the engine in situ.**

Engine stand
I believe that to build these engines successfully a proper engine stand is essential. Universal engine stands are available from many sources quite cheaply. However, for the Lotus twin cam they all have the same fault in that they connect to the engine by the bell housing fixing bolts which prevents any work at the rear of the engine including fitting of rear seal, flywheel, clutch and if you cannot fit the rear seal you cannot fit the sump. It is better to modify the fixing with a specially made bracket to enable bolting to the left engine mounting support boss. Make sure that the bracket is long enough so the engine can fully revolve on the stand and make sure that it is a properly made bracket, it has to take a good bit of weight. If you have any early photographs of Lotus building twin cams at Hethel you will note that this is how they used to do it.

Cleanliness
No engine wants dirty internals. Fast revolving machinery is going to wear and break much quicker with dirt and grit mixing with the lubricant which will carry it round to the bearings. When an engine is machined the machine shop should always be asked to clean out the oil and water galleries. Some machine shops do this as standard but it is better to ask and make sure. Your new engine oil will now be clean and totally free from all sorts of muck in the galleries. Your cooling water will flow to all parts of the engine. This obviously is an important step in the overall cooling of the engine. The machinist should have also cleaned the crankcase after machining to remove all those tiny bits of metal swarf that can do so much damage. It is a good idea to clean again just to make absolutely sure. If you have not got a compressed air line available please do not clean up with dirty rags, use pristine clean ones. Clean all old gaskets from the crankcase before assembly starts.

Core/freeze plugs
In any engine rebuild these plugs should always be replaced. With a full machining job and cleaning of the oil and water galleries the old ones are removed anyway. They may look super on the outside but they corrode from the inside out and although they look good they may be paper thin. The way to remove core plugs is to knock inwards on one side so the plug twists, then pull out edgeways with a pair of pliers. Tap in new with a drift with a covering of Wellseal. New plugs are 1 5/8 inches diameter. Sometimes the rear one in the bell housing is larger.

Bearing lubrication
Bearing surfaces should be lubricated as they are installed in the engine, in fact all moving surfaces should be lubricated. It is quite usual for a rebuilt engine not to be installed or run in a car for many months and if normal engine oil is used this will gradually drain away leaving the bearing surfaces dry. It is not good news for an engine to be started with dry bearings.

I use Graphogen to lubricate the bearings, a black grease specially formulated to protect bearing surfaces during the initial start up but there are several other assembly lubricants available. Whatever you use it must stick to the revolving surfaces but be easily soluble with normal engine oil.

Checking Rotation

When assembling the engine it is important to check the ease of rotation of moving parts after the tightening of each new bearing shell. If there is a binding problem you will know then exactly where the problem lies.

Gaskets and Sealant

Good gasket sealant is a basic requirement to keep the oil drips away from the twin cam and Wellseal is the one I use together with Reinzosil silicone in specific areas in carefully controlled amounts. If used indiscriminately little bits of squeezed out silicone will be going round your engine, blocking up oil ways. Who said?: 'Although silicone is good, twice as much does not make it twice as good', a wise man. When using any gasket sealant cleanliness is essential. Wiping the surfaces with an old oily rag is a nonsense because any trace of oil on the surface will prevent the sealant adhering to the surface and thus sealing. In an effort to cure the timing chest oil leak problem in my daily road going engine for the last 400,000 miles I have made up gaskets for the timing chest and these have been most successful. I must also stress that the standard use of a torque wrench to get all the nuts and bolts tightened equally to the correct torque cannot but help the oil leaking problem and must be correct engineering. Lotus and other car manufacturers do not make up Torque Tables just for fun.

Warning

If you turn any part of the engine over before the timing chain is fitted and without due consideration of the other rotating parts, the valves could hit each other or the pistons and cause damage.

Bearing Caps

Before taking any engine apart it is essential and good engineering practise to mark bearing caps to ensure they go to the correct main journal or connecting rod. 1 No. centre pop for Number 1 mains journal for instance.

The inlet and exhaust cam caps are factory numbered but must be fitted the correct way round. See photographs. There can be confusion over caps '6' and '9' but if fitted the correct way round as the photographs show there should not be a problem.

Piston Rings

If rebuilding an engine without a rebore the cylinders should be honed to take out the glaze and to enable new rings to bed into the bores during running in. It must always be correct engineering to fit new piston rings. It is advisable to check the ring gap. Insert the ring into a bore and push down about 2 inches with an inverted piston to ensure the ring is square in the bore. Measure the gap. Lotus quote 9 to 14 thou. for compression rings and 10 to 20 thou. for oil control rings. It is advisable to aim for the lower figure as wear will take place. Generally as a rough guide the ring gap in thou. of an inch, should be about 3 times the diameter of the piston in inches........ so, for example 3 x 3.25 equals just under 10thou. and is what we are aiming for.

REBUILDING THE ENGINE CYLINDER HEAD

General
Most of the restoration jobs on the cylinder head are professional only tasks simply because of the machinery required to do the job. Replacement and grinding in of a new valve(s) can be done by the owner mechanic as can the necessary tappet clearance shimming. Even so, without a micrometer and without a box full of shims the clearances job can be almost impossible. You may not be sure of the existing shim size. You telephone a supplier for shims of a certain size only to find when checking the resulting clearance that the calculated shim is not quite right and you need to start the whole exercise again. A micrometer is essential for measuring shim thickness, yes new ones are etched with the size but this etching wears off especially if the shim is installed with the etching upwards towards the tappet.

A full head restoration can be expensive and it is a good idea to obtain a quote from your preferred machinist. Before any machining takes place the head has to be stripped of cams, valves and studs. If you have a valve spring compressor and a stud extractor you can save yourself a few pounds doing this job yourself. If the rebuild is only a valve trim up, then the valves, tappet buckets and cam bearings should all be identified so they go back in the head where they came from. New tappet buckets should be identified as the sleeves are honed individually to each tappet. See photograph. A permanent marker on the face is the way I do it and, for example, 1EX or 1IN tells you exactly where it goes.

When removing studs check that the aluminium thread is sound in the head. These 5/16 UNC threads especially for the exhaust manifold are prone to stripping. A stainless steel insert is easily installed and is stronger than the original. The plugs have been screwed in and out a few times and the 14mm thread can be very worn. The same type of stainless steel insert can be installed very easily and cheaply. Some owners may even have the simple tooling to do it themselves. A half hour job. This job can also be done with the head in situ on an in-commission engine although great care is required to prevent swarf entering the combustion chamber.

These heads are getting old and in most cases have done a few miles and therefore it is a good idea before the machining starts to check for cracks.. The standard crack is between the plug hole and the valve seats especially so on the Big Valve engine. The waterways on the head face can be corroded. Modern aluminium welding techniques make these repairs a standard job on the twin cam and when the repaired head is returned to you no evidence of the repair can usually be seen. With waterways welding the head obviously has to be skimmed and it is a good idea to insist on a minimum skim. The standard head should be 4.64 inches depth and the Big Valve 4.60 inches. Over 40/50 years some heads are beginning to get a little 'thin' and care against over skimming should be taken. Every skim off the head increases the compression ratio.

Valve Seats
The valve seats wear over time and the valve clearances have to be adjusted. A head with new seats and new valves should have shim thicknesses of around 120/130 thou. Worn seats at the end of their life will have thin shims and I do not like to go much below 70 thou. Any less than this minimum thickness can cause the shim to break up. The shim thickness therefore is a good indicator when new seats are required. The old

seat is a machined out and the new one 'pressed' in. Pressed may not be the correct word because with the head heated to 150 degrees Centigrade and the seat cooled in a freezer they do go in without a problem.

Valve Guides

No valve stem seals are fitted to the engine therefore the fit of the valve to the guide is vitally important. Smoke from the exhaust is a good indicator of oil in the combustion chamber and one of the reasons for this is oil draining down the valve guide from the head and helped along with worn tappet followers. It is a good idea to fit new valves with the new seats and it would be silly to install the valves into worn guides.

Guides are knocked out with a specialist drift with the head heated to 150deg. centigrade. The drift has an extension to fit into the guide therefore keeping the drift securely in place. New guides are installed much the same. Both inlet and exhaust guides have a circlip and the guide is knocked in until the circlip is on its seat. Care is needed here because excessive force with the circlip on its seat can dislodge it. The guide needs to be reamed and honed to the valve stem.

With the new guides installed now is the time for the new seat to be machined for the valve angle to ensure concentricity and cut as necessary to obtain the correct height for new clearances.

Unleaded Petrol

All seats today will be suitable (or should be) for unleaded petrol so valve seat recession should not be a problem. I do not and have never used a fuel additive in any of my Elans and bearing in mind the mileage I do (15000 annually) I have never had a problem. For road going twin cams cast iron guides should be used, bronze only really necessary for high engine revs. and racing.

Valve Grinding

Whether replacing a single burnt out valve, just trimming up all valves or grinding in a full set of new valves the technique is the same. You lap the valve in first with a coarse paste then with a smooth paste to get a uniform dull but flat finish all round the valve and seat. It is essential to keep the grinding paste away from the guide. When happy with the seat and valve it is essential to clean all the grinding paste, swarf, and muck away from the engine with a copious amount of cleaning fluid. A compressed air line helps here but if not available very clean cloths.

With well worn valves a valve grinding machine is essential which does entail a visit to your machinist. The reground valve will still need a light lapping to the seat which in all probability has also been cut. This operation is where you must be aware that you can do the work and then find the valve required shim clearance is too small.

Tappet Followers (Buckets)

On the early engines the tappet buckets ran directly in the aluminium. Experience showed with the aluminium wearing and allowing oil to drop down onto the valve stem, any wear in the guide produced a very smokey engine. Lotus therefore installed a cast iron sleeve in the head, first on the exhaust side being hottest and then also on the inlet side to deal with this problem. These sleeves can be installed retrospectively. New sleeves should be and honed to the new tappet bucket.

Cutting Out Old Seats
The old valve seat is machined out using the valve guide to keep the cutter concentric to the seat. The very thin portion of seat left virtually falls out leaving the originally machined recess ready to accept the new seat.

Head Recesses
All ready for the new valve seats

Below
Pressing in New valve Seats.
The head is heated and the seat chilled in a freezer and then comparatively easily pressed into the head. The seat is machined for the valve in situ again using the guide to obtained concentricity.

HONING

This is a silicon carbide hone that has just been used on a tappet sleeve. It is important that new sleeves are honed to size for the tappet bucket. If very tight as this one was the bucket can seize holding the valve down just ready to say "hello" to the piston. The hone is used in an electric drill and in conjunction with cutting fluid. Hone a little at a time and test frequently for bucket fit.

The hone for a valve guide is very similar except obviously thinner and longer. You are looking for a good sliding fit but without play

TAPPET SLEEVE
This is a tappet sleeve which has been retro fitted to an early head. See also the nice new valve guide

TAPPET BUCKET
A tappet bucket. A good fit from honing is essential to keep excess oil away from the valve guide. With the cam lobe slightly offset the bucket revolves in the sleeve

VALVE GEAR (Above)

Spring seat, outer and inner valve springs, retainer. These items are common to all 4 inlet and 4 exhaust valves. 2 springs are used to allow the twin cam to rev. freely up to 6500 revs, without valve bounce.

VALVE (Below)

An inlet valve with its 2 collets

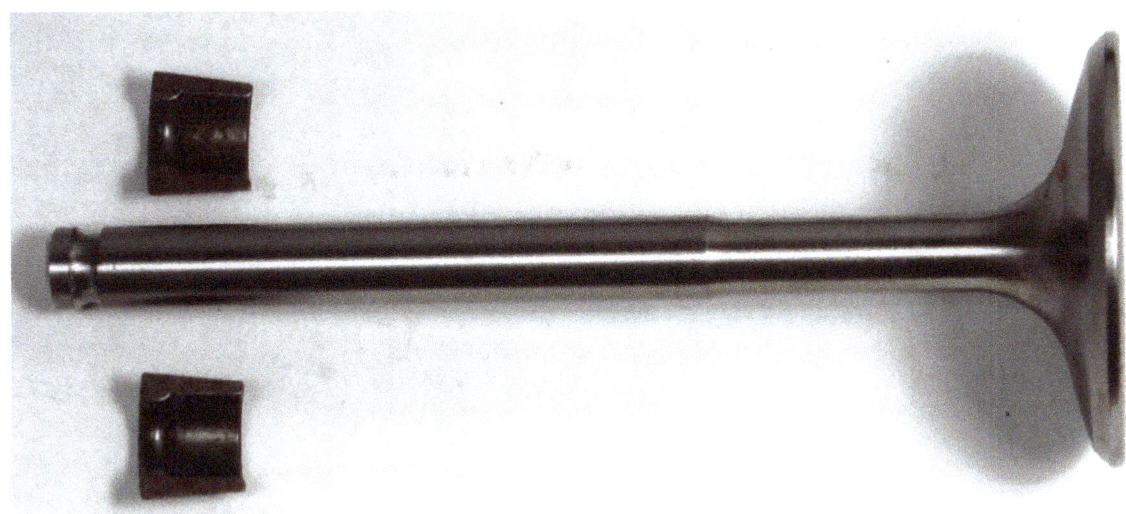

VALVE COMPRESSOR (Below)

A standard valve compressor for overhead cam engines.
I have been using this for some years

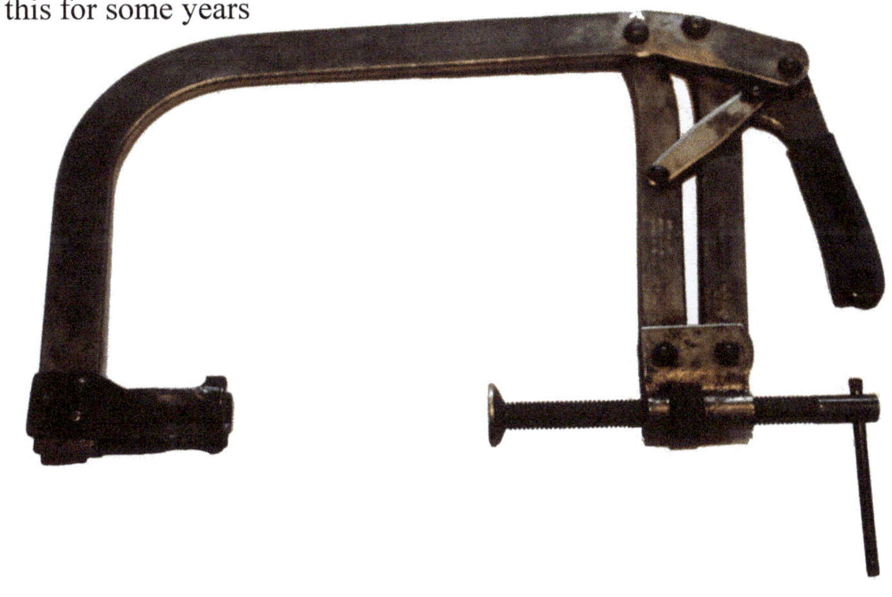

CAM CAPS.

(Left) The cam caps are in line bored so every one is unique to the head that it belongs to. This is why not only the cap number is identified with number stamped on it but also the head, in this case AA. Some heads, especially early ones, are not identified on the cap and it is imperative that these are not mixed up with other cam caps.

When installing cams it is essential that you spin the cams over with a socket on the sprocket retaining nut to check that there is no binding.

(Below) The corresponding head for the cap photographed showing the head identification number.

CAM SHAFT BEARINGS.

A full set of new cam bearings, 20 No. in the set. 16 of the bearings are full 0.75 inches width but 4 No. are only 0.5 inches wide. The narrow ones go to cam cap numbers 1 and 10. All cam cap nuts should be torqued to 9lb. ft. Tighten down all nuts in equal steps. The nuts on the cap studs which are longer to secure the cam cover can be tightened and torqued with a long reach 1/2 ins. AF socket.

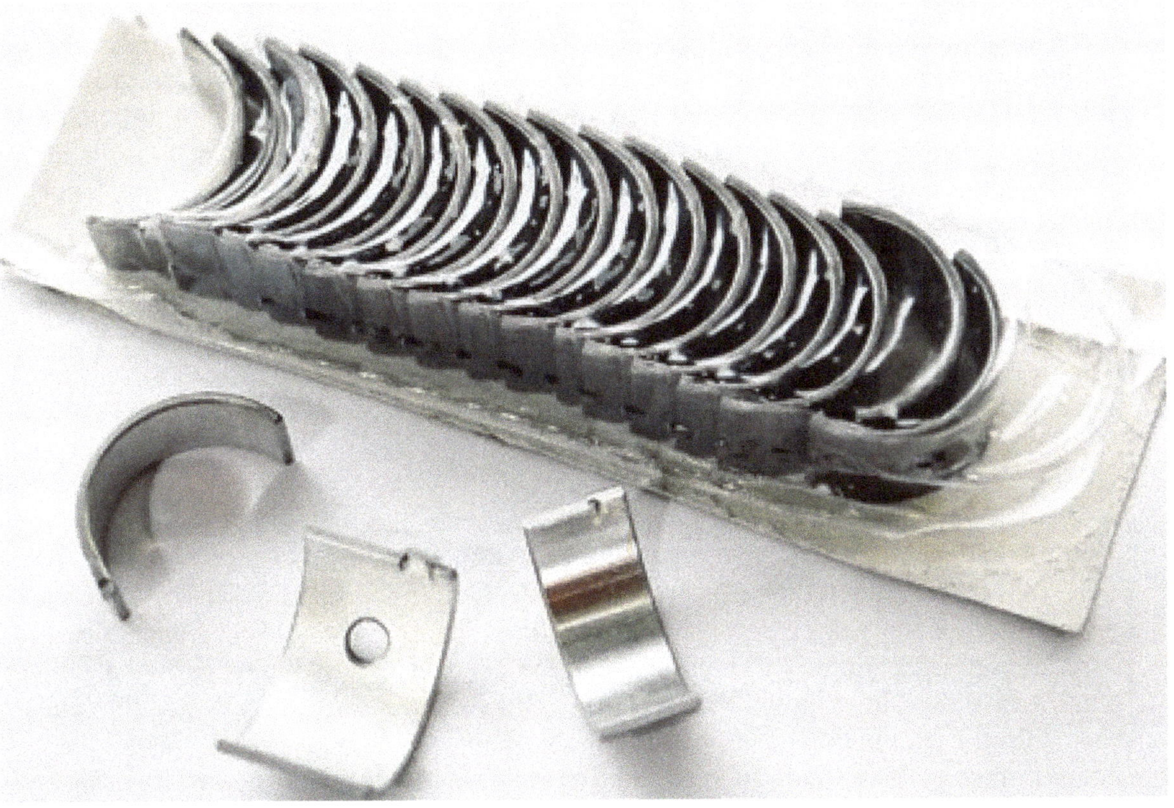

VALVE GUIDES (below)
Inlet and exhaust valve guides of bronze and cast iron. Inlets are the bottom ones with the tapered end to give less obstruction to air flow.
Bronze guides are really for racing, giving better lubrication at high revs. Cast iron are for road going engines.

The operating cam lobe is slightly offset from the bucket centre which generates a turning motion to the bucket. This turning in operation keeps wear here to a minimum not only keeping excess oil away from the valve stem but also keeping tappet clearance wear very much in check. Before unleaded petrol came into being the engine could do 10,000miles before a check on valve clearances was necessary. With valve seat recession added into the equation today it's a good idea to check every 6000 miles and certainly after 500 miles with a new engine.

Rebuilding the head
The head can now be rebuilt. Install the valves with new springs. It is a good idea to give the installed valve a couple of taps with a drift on the valve stem to ensure that the collets are properly seated. Install the studs with a stud extractor/installer. I torque up to the relevant tightness for that size thread, 12-15 lb.ft. for 5/16 UNC, with a touch of Loctite.

Install shims with thickness relevant to what has been done to the head. With just a valve trim I insert shims of about 20 thou. less than came out. With new seats and valves, shims between 80-90 thou. should give you a clearance to measure. Install the tappet buckets with a touch of Graphogen.
Install new cam bearings in the head and cam caps. The half size shells go to cam cap positions 1 and 10. Do not forget the lubricant.
Install the first cam using standard nuts on the cam caps not the final Nyloc nuts. When installing any cam I always have the lobe on No. 4 facing inwards. The valves will not open then as you tighten the cam caps. The nuts should be torqued to 9lb.ft. When the cam is installed check for ease of turning and that it is not binding. When each lobe has passed its full open point you should be able to feel the valve spring pushing the cam round. Leave the cam with the lobe on No.4 cylinder facing inwards towards the plug well. With the cam in this position all valves will near enough be closed. Install the second cam exactly as the first. With the first cam having its valves closed it is safe to turn the second cam.

Valve Clearances
The valve clearances should be;
Inlet.......... 5-7 thou. calculate for 6 thou.
Exhaust..... 9-11 thou. calculate for 10 thou.

The original exhaust clearances of 6-8 thou. as stated in the original Lotus Workshop Manual, have been superseded in the light of experience and different valve material and should not be used. If you aim for the average figure of 6 thou. inlet and 10 thou. for the exhaust you will not go far wrong. To do these clearances properly can take some time even if you have done it a few times before. A look at my last calculation sheet should be of interest to those 'first timers'.

Turn a cam so that the lobe is pointing directly away from the tappet and the cam base circle is pointing towards the tappet. Measure the clearance carefully with feeler gauges and note the reading. Measure the other 3 clearances on that cam the same way. I do one cam at a time so I do not get into trouble turning a cam when I should not. It is simple addition and subtraction arithmetic to calculate the correct shim needed.

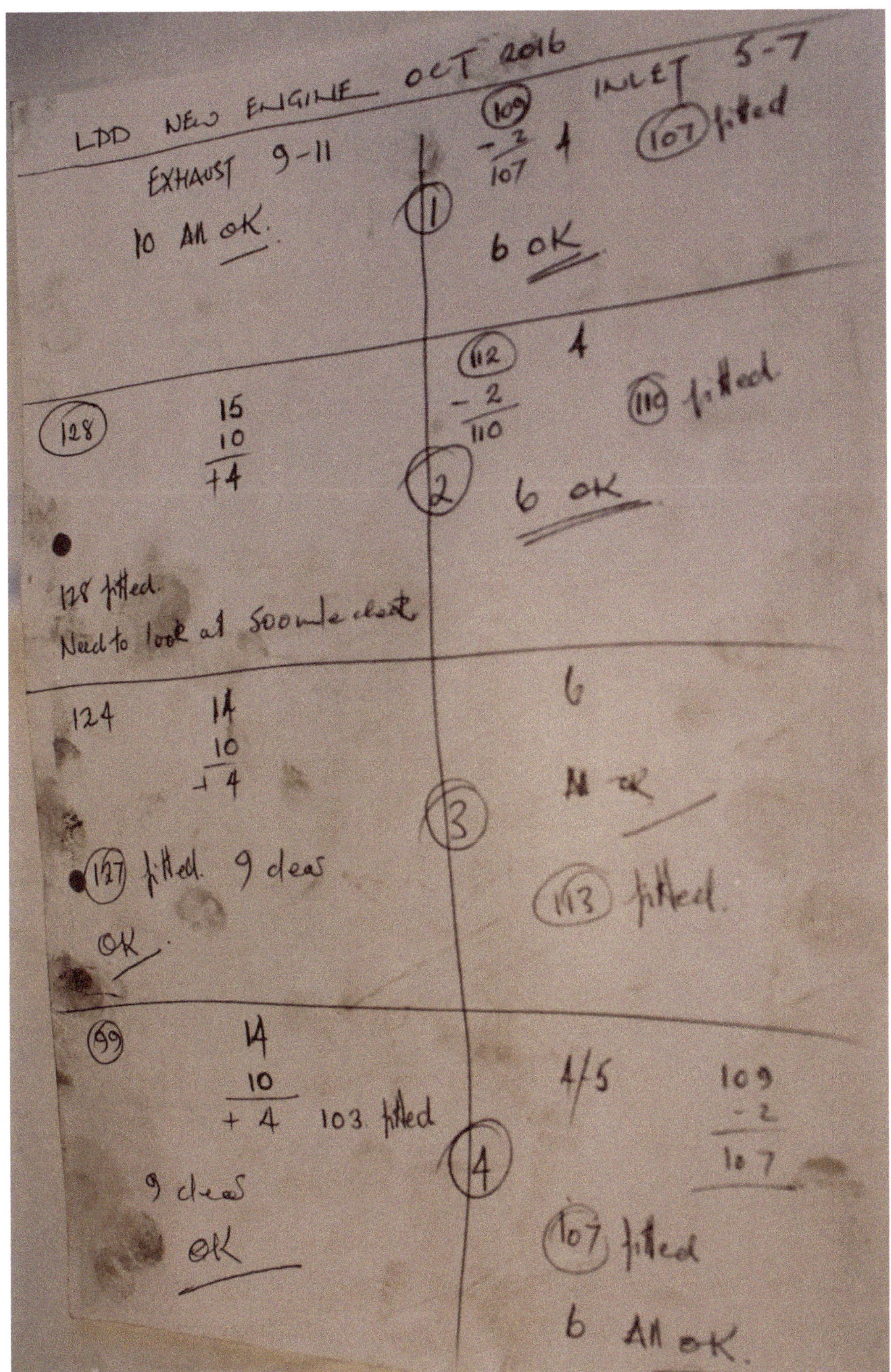

TAPPET CLEARANCES

Checking clearances on my own twin cam. I line out an A4 sheet with 8 sections for each calculation of the new shim size. If you add to the shim size you decrease the clearance and if you reduce the shim size you increase the clearance. Note the exhaust valve calculation for number 2 cylinder. I left it as is because unusually for me I did not have correct shims in stock. See also my note to sort at 500 mile check. I do not mind too much clearances slightly on the high side after trimming valves, with bedding down clearances usually reduce anyway.

Looking at one of my calculations say inlet valve on No. 1 cylinder;

Measured clearance 4thou. but required clearance is 6 so the shim must be DECREASED in thickness by 2 thou. The original shim was 109 thou. so the required new shim should be 109 minus 2 giving a shim of 107 thou. This shim was fitted and the clearance checked which was found to be correct at 6 thou. The cam might have to be removed and installed a few times to get all clearances correct. When fully happy the standard nuts can be replaced with new Nylocs. Ensure that you leave the cam with the lobe on No. 4 cylinder pointing inwards so that you can turn the second cam without a problem.

The head is now complete and ready to be mated to the crankcase.

REBUILDING THE TIMING CHEST AND THE WATER PUMP.

Removal from the Crankcase.
Removing the old water pump.
Pulling the timing chest away from the backplate and crankcase can be difficult especially if it has not been removed for some time. The pump 'O' ring insert can be well stuck into the back plate. The back plate is only held onto the crankcase with one ¼ inch UNC bolt and if not careful the plate can bend and distort making water and oil sealing difficult during re-assembly. For difficult removals I always clamp the back plate to the crankcase to prevent this and with a solid base, removal is easier. Please see photograph.

Using a puller remove the water pump drive hub. Remove the bearing shaft location clip. Push out the pump bearing. Heating the chest in boiling water makes this much easier. Remove the 'O' ring insert. A flat plate JUST small enough to go into the insert at an angle, then centralised, makes the right platform to push out the insert.

Thoroughly clean the chest, back plate and 'O' ring insert. Check the screw threads for stripping especially the sump connections.

Ensure that the water drainage hole from the bearing housing in the timing chest is clear.
Ensure that the oil jet hole for the timing chain is clear in the back plate.

Oil Dip Stick Pipe
Check that the oil level dip stick pipe is exactly 4.1 inches above the sump flange. If not or the pipe has a damaged flare at the top, replace the pipe with a new one. Heat the chest by immersion in boiling water, knock out the old pipe. Having put the new pipe in the freezer, and heated the chest in boiling water again, insert a 5/16 bolt into the new pipe to prevent damage to the flare and the pipe should now tap in reasonably easily to the 4.1 inches dimension. This is MOST important for correct oil level. If you have a flaring tool it is possible to reuse the original tube after making a new flare. The pipe can split if not heated when the tool 'flares' out the pipe.

Water Pump
Heat the timing chest again in boiling water, press in the new pump bearing from the inside of the chest and pressing on the outer sleeve of the bearing, not the shaft, until the retaining clip grooves line up. The short spindle of the bearing faces forward. Pressing the pump bearing in from the inside of the chest gives the timing chest better support, the timing chest boss itself being directly supported. Pressing on the pump bearing housing prevents massive stress on the bearing which can lead to early failure. Insert the retaining clip.

Supporting the shaft, press on the drive hub until the face is flush with the shaft.

The modern pump kit has a steel cup that presses into the timing chest bearing housing so with a smear of silicon on both surfaces and with the bearing being supported on the drive hub press the cup into the chest.

REMOVING TIMING CHEST.

Clamp the back plate to the block or in stubborn cases you will bend the back plate, it is only held on with 1 No. 1/4 UNC bolt. If the back plate bends oil leakage can only get worse.

I put a hard wood half moon plug in the cylinder bore to prevent damage from the G clamp. The clamping face of the wood plug is angled so that the G clamp clears the engine block.

Two mole grips at the bottom complete the job and not only does this prevent damage to the back plate it also makes the job so much easier. The timing chest has already been removed for clarity.

A hand at the top, one at the bottom, a bit of a wiggle and off the timing chest comes with no problem.

It is also so much easier with an engine stand.

Taking care not to damage the pump seal face locate the seal over the bearing shaft and into the cup with some silicone and press in.

Fit a new 'O' ring on the insert housing and using a liberal amount of silicone round the outside push into the timing chest. Take no notice of the Lotus Manual which says install dry! The locating lug goes into the slot in the timing chest.

Install the other half of the pump seal onto the bearing shaft ensuring that the seal is the right way round. The rubber part installs directly next to the impellor.
Supporting the bearing on the centre of the drive shaft hub, press on the pump impeller. Do not press on the edge but on the central boss especially with a fabricated steel impeller. A ½ inch socket neatly fits on this boss. Clearance between the impeller blades and the insert 'O' ring housing should be 25 / 30 thou.

Check rotation of the pump.

Screw the timing chain damper to the chest. Use Reinzosil on the 2 ¼ UNC screw threads to prevent oil leakage. If the damper strip is well grooved install a new one.

Push in a new crankshaft seal using a little Reinzosil to ensure a good seal.

The timing chest complete with new pump is now ready to install onto the engine, together with the back plate.

MODULE WATER PUMP
Today an improvement to the timing chest is available with the water pump enclosed in a module that can be withdrawn from the chest and replaced without all the drama and bank balance needed with the original design. This is a super modification and I am trying one out on my own high mileage Elan at the present time.

Please see the photograph of my Elan engine without the pump module. To my knowledge two designs are available and I believe a third is imminent. If you need a new timing chest anyway which would be a good proportion of the cost, this would be the way to go.

WORK ON THE TIMING CHEST

Oil Dip Stick Tube Replacement.

The engine oil dip stick pipe wears after long use on the flare of the top edge allowing the dip stick to give a false higher reading. Fitting a new pipe is easy. Immerse the chest in boiling water, freeze the oil pipe and tap into the hole. Use a bolt into the 5/16 pipe to prevent damage to the flare when tapping in. The pipe top should be 4.1 inches from the bottom flange where the sump goes.

You can also see the timing marks in this photograph, TDC, 10, 20 & 30 degrees before TDC.

Timing Chest Crankshaft Oil Seal (below)

The rubber oil seal can be tapped out and a new one tapped in. I say tapped advisedly as it does not take much force to ease these seals in. Smear a layer of silicone round the aluminium housing and on the rubber seal. Push the seal initially into the housing with fingers and then using the reverse side of a bearing insertion tool, sized to be just smaller than the housing, carefully tap the seal in till flush with the timing chest. Clean off the squeezed out silicone.

TIMING CHEST WITH REMOVABLE MODULE WATER PUMP
A timing chest from Dave Bean in America showing the design of the chest with a removable water pump module. Apart from installation of the module there is no difference in assembly. However there is considerable difference in ease of repair in the event of pump failure.

This timing chest was installed in my engine in September 2016 and has done approaching 30,000 miles.

WATER PUMP MODULE

The module holding the water pump is machined from the solid and is a super bit of kit. The original pump has a 1/2 inch bearing but this is a size larger to give even better reliability. I did enquire about the failure rate. NONE I was told.

Sealing to separate oil and water is exactly as original and the O rings can plainly be seen. The back plate although a very good fit round the O rings and tighter than original I would still re - commend a coating of silicone.

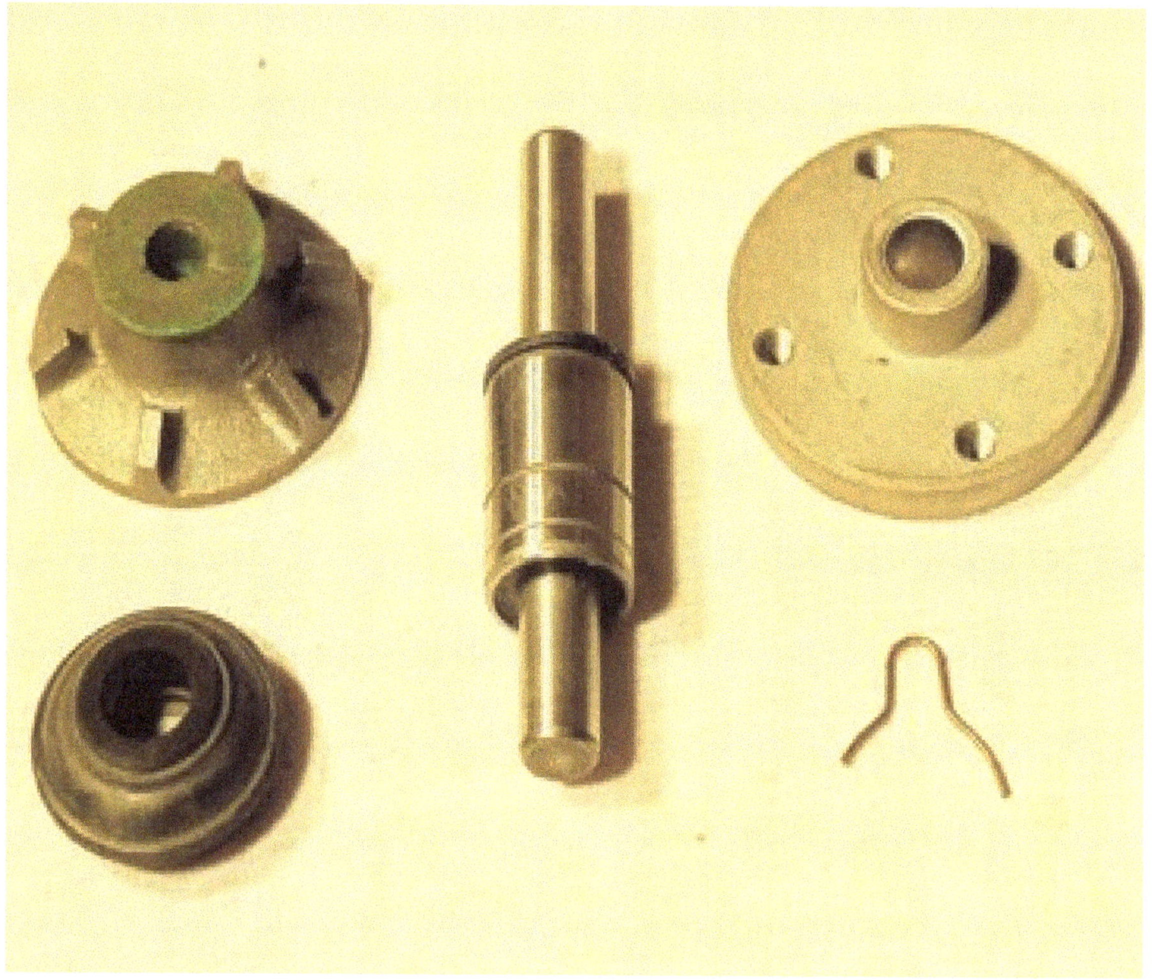

WATER PUMP COMPONENTS.

The water pump has to be pressed into the timing chest, an easy job to do if you have a hydraulic press. The installation procedure is described in the following photographs and text. The water pump is one of the Achilles heels of the car.

DRIVE PULLEY AND BELT.

Just a reminder photograph really. The water pump is a weak point on the twin cam engine and extra load on the water pump bearing is likely to cause an early failure. It is essential that the drive belt is not over tensioned so please take heed of the Lotus stated 1/2 inch movement. If anything I run mine a shade slacker than that.

INSTALLING WATER PUMP.

1. Bearing.
Heat the timing chest in boiling water.

Press the bearing into the timing chest, pressing on the outer case of the bearing not the shaft. This prevents massive side loads which will severely reduce the life of the bearing. Although not clear from this photograph, there is a tube over the shaft and directly pressing on the bearing case.

When the 2 No. retaining clip grooves line up, remove from the press and install the retaining clip.

INSTALLING WATER PUMP

2. Fan Pulley Boss

Support the bearing shaft. If you support the timing chest the bearing is being stressed and could move in the chest.

Press on the pulley boss until flush with the shaft

INSTALLING WATER PUMP (cont.)

3. Water Seal.
Smear some silicone round the housing and the seal and push the seal into the timing chest taking great care not to damage the impeller running surface.

4. '0' Ring Insert
The O ring insert shown with one O ring fitted and the other loose. Always use new O rings. Lotus say install dry which would be a mistake. Always coat with silicone. You can plainly see the lug which locates the insert in the timing chest.

INSTALLING WATER PUMP. (cont.)

5. Fitting Insert.

Fit new 'O' ring round insert inner groove, smear with silicone and push into the timing chest. No need to force. There is a lug on the insert which locates in a recess in the chest. In the picture you can just see a line of silicone round the bottom of the insert where the assembly has been wiped clean.

Note. Please see the left hand fixing bolt hole and the boss cut away by the timing chain. This is not bad design it is bad maintenance not keeping the chain at the correct tension.

6. Impeller.

Oil the bearing spindle and supporting the spindle this time, press on the impeller. It is easier if it has been soaked in boiling water. The clearance between insert and impeller should be 25/30 thou. Please note feeler gauges. This photograph has not been 'set up', this timing chest is destined for a real engine.

This is a cast impeller. Some impellers are fabricated and are not as robust as the proper job. You have to be very careful during installation that you do not distort them.

A COMPLETED ENGINE REBUILD

The completed engine all as described in the following text and as shown in the illustrations. If I was doing a real 'Concours' job I would have painted individual items Lotus grey before assembly. However this is a working engine and the paint does not make it run any better.

The petrol pump, oil pump and distributor have to be added before rotating the engine on its side for the flywheel and clutch assembly.

On the other side of the engine I do not fit the alternator before installing in the car as it will get in the way of fitting the exhaust.

ASSEMBLY OF THE ENGINE

1. Jackshaft Bearings
These bearings are not highly stressed as the original job of the Lotus jackshaft was a cam shaft in the push rod Ford engine. Obviously, it no longer carries out this function on the "twin cam" and it is possible, therefore, that these bearings do not always need to be replaced in a rebuild. However, please check carefully for any play between the jackshaft and the bearings as on some rebuilds they will have to be replaced. Importantly you must note that all three bearing shells are different. The front and rear shells are 0.75 inches long and the central one 0.64 inches long. The front bearing shell has 2 oil way holes, one in from the oil gallery and one out that runs to the top of the block for the lubrication of the cams in the head. This shell also has to be installed the correct way round as the oil holes are off set. The installation of these bearing shells can be tricky especially the central one. Lotus show a puller in their workshop manual that is used with a dolly inserted in the bearing shell. This puller is obviously a Ford original tool and is even called a 'cam' bearing bush remover / replacer instead of reference to the jackshaft. I use a simple made up drift that not only will remove shells but also replace them. See photograph. The outside diameter is a good but sliding fit in the crankcase bearing boss which keeps the drift square to the central bearing boss. The bearing shell must be inserted in the correct position for the oil holes to line up. I mark the crankcase and the bearing shell outer case with a dab of white paint. The shells are tight and do require some force to install and this can cause slight distortion of the shell especially when knocking them in and it is very normal for the jackshaft to be tight and difficult to turn. The shells will need honing on high spots. On the centre bearing to reach it I use the hone on a flexible drive. Check frequently as the shaft very quickly will free from binding. Make absolutely sure that the bearings are clean from any swarf when finished. In my early days before I had all the correct tools I used a very fine emery cloth (and even scraped smoother) stuck to a round dolly and very gently eased it round the bearings. This was successful but not as good as a professional hone.

Your machinist will insert these bearings for you if required and it is recommended that this is probably the better way to go for the average owner mechanic.

2. The central bearing is installed first. Insert the drift in the block, put the bearing shell on the end, line up to the paint marks and gently knock the shell in till the oil holes line up. On the difficult to see centre bearing, a good check can be made on the oil holes alignment with a bent length of welding rod. The 2 outer shells are far less of a problem but the same technique applies.

3. Jackshaft
Apply lubricant to the bearing shells and jackshaft and insert the shaft into the block. Fit the shaft thrust plate and locking washer, torque up the ¼ UNC bolts to 6 lb. ft and bend over the locking tabs. Check the shaft for easy rotation.

4. Crankshaft
If your crankshaft has been re-ground which is normal today, it is 50 years old, make absolutely sure you have the correct oversize bearing shells for the mains and big end journals. If you install incorrect bearings you will have no oil pressure and all your careful hard work will be for nothing.

Insert the main bearing shells into the crankcase and into the bearing caps. Apply lubricant to the bearing shells and the crankshaft main journals and gently lay the shaft into the block.

5. The end float of between 3 and 8 thou. is controlled by thrust washers housed in the centre bearing web. They are held in place by the centre bearing cap. End float can be measured either by feeler gauges or by a dial gauge on the rear flywheel boss. Thrust washers come in several sizes from standard to plus 15 thou. in approximately 3 thou. steps. Insert the thrust washers pushing them round the slot between the shaft and the centre web. Importantly the oil grooves face outwards towards the rotating shaft not the stationary web. Push the shaft hard one way and measure the float. Normally your machinist will sort out what thrust washers you need.

6. Making sure the main caps go to the correct bearing, install the bearing caps and torque up to 60lb.ft. in 2 stages. Check the shaft for rotation after tightening each cap.

Piston, Piston Rings and Connecting Rods
7. Taking careful note of the pistons and connecting rods front, apply lubrication to the gudgeon or wrist pin and push into each piston and conrod small end. Install the pin securing circlips. Make absolutely sure the clips are secure in their locating groove.

8. Hold the con rod loosely in a vice with the piston resting on the vice jaws. The rings have to be installed in reverse order, the bottom oil control ring first, then the 2nd compression ring, then the top or first compression ring. Having checked the ring gap in the relevant bore, expand the ring with piston ring pliers and making sure that the ring is the correct way up, gently ease it over the piston into its correct slot. Normally the ring is etched on the top edge with the manufacturers' part number and to identify the top edge for you. Fit all 3 rings and position them so that the 3 gaps are roughly 120 degrees apart.

 9. Insert the big end bearing shell in the con rod. Lubricate the bearing shell, the piston and bore. Clamp the ring compressor tightly to the piston and making sure the piston is facing the right way and the conrod is in the right bore, insert the piston into the bore. Tap the piston fully into the bore with the wooden end of a mallet. With the crankshaft in the BDC position guide the conrod onto the bearing journal. Insert the bearing shell into cap and lubricate before installing on the conrod. You cannot install the wrong way. Insert the bolts and torque up in 2 stages to 45lb.ft. Check crankshaft for ease of rotation.
Note.
Make sure the ring compressor is not upside down. Normally they have indents on the bottom edge preventing the compressor trying to enter the cylinder bore.

10. Repeat stages 7, 8 and 9 for the other 3 pistons.

For the installation of the timing chest I turn the engine to the vertical position on the engine stand. I also remove the timing chain adjustment plunger mechanism as this makes the installation of the chain and timing of the cams easier.

11. Apply a thin coat of Wellseal to the front of the crankcase and the rear face of the back plate. Lay the paper gasket in position and install the back plate onto the crankcase. There is only one ¼ UNC fixing bolt and the temporary installation of 2 additional timing chest bolts guarantees the correct position before tightening the one fixing bolt to 6lb.ft.

12. Install the jackshaft sprocket not forgetting the locking washer. Bend the locking washer tabs onto the bolt heads. Dress neatly with a small hammer.

13. Lay in the timing chain. Always use a new chain the expense for this item is minimal.

14. Slide the oil slinger onto the crankshaft

15. (With Gaskets)
Coat the 2 outside edges of the back plate and timing chest with Wellseal. Lay the gaskets on the back plate.
Smear Reinzosil round the timing chest water pump boss and lay on the pump gasket. Slide the O ring onto the insert. Smear silicon round the O ring, gasket and round the pump boss on the back plate.
Push the timing chest onto the back plate, the pump insert going into the back plate boss.

Note.
If you prefer to install the timing chest with no gaskets as the original engine the best way is with Reinzosil very slightly domed on both faces. Let cure before installation. This to me is a gasket anyway!!

16. Install all the fixing bolts tightening all bolts to 6lb.ft with the torque wrench. With the exception of two 5/16 diameter bolts, one round the pump boss and the bolt holding the belt tensioner, all fixings are ¼ inch diameter but ALL should be tightened equally to 6lb.ft

Crankshaft Pulley
17. Install the crankshaft pulley. Grease the shaft to prevent the oil seal running dry when first starting. It is a good idea to paint the notch on the inner flange which indicates TDC, white, to enable the timing strobe to pick up the mark more easily when checking dynamic ignition timing.

Cylinder Head
18. Turn the engine so that the pistons are in mid position in the bores with the piston in No.1 bore on the up stroke if you turn the engine in the normal direction, clockwise.

On the cylinder head ensure that the cam lobes on No.4 cylinder are pointing towards each other which should mean that for practical purposes all valves are closed and the valves are positioned for TDC No. 1 cylinder. If still in the head remove the timing chain tension quadrant and in the timing chest remove the tension adjusting mechanism complete.

Ensure that the cylinder head bolt threads are clean and lubricate with a little Copper Slip.

19. Smear Wellseal on top of the timing chest and on the bottom of the head where the timing chest connects to. Lay the cork gasket on the chest.

20. Place the head gasket in position. It will only go one way with the shape of the block and head but the copper side is uppermost.

21. Screw in the head locating studs/guides. I use head bolt positions 7 and 8 on the Lotus tightening sequence drawing for these to remind me this is where the special washers with flats are installed. The flats are there to allow clearance to the No.1 inlet camshaft cap (head bolt 7) and clearance to the exhaust camshaft float controlling boss (head bolt 8)

22. Lay 5/16 diameter spacer rods on each end of the head gasket to prevent the head going fully down.

23. Smear Reinzosil on the inside and round the edge of the breather hole in the crankcase. Smear Reinzosil round the top flange and bottom of the rubber breather pipe and insert securely into the breather hole. Smear Reinzosil on the cylinder head connection.

24. Line up the head with the guides and lower carefully onto the spacers. Screw loosely in the cylinder head bolts making sure you have the correct washers. Screw in loosely the 3 bolts for the timing chest.

25. Make sure the rubber breather pipe is exactly lined up with the head, take the weight of the head and ease out the spacers. The head should be now down on the crankcase in exactly the correct position. Check that the breather pipe has installed correctly to the head. Remove the guides and install the final 2 No. head bolts in their place with the correct washers to miss the No. 1 cam cap and the exhaust cam end float flange.

26. Torque down the head to 65lb.ft in 3 stages, 35, 50 then 65lb.ft. in the correct sequence as the Lotus illustration drawing. You may have to remove No 1 cam cap to get the 9/16 socket on the adjacent head bolt. The special washers will try to turn as you tighten their head bolt and you will have to hold them with long nose pliers until you get some tension on them.
Note.
It is an idea to keep a good check at the timing chest cork gasket as you begin to tighten the head bolts and keep it pressed in place with your fingers. There is a tendency for this gasket to spread outwards and this is an extra bit of finesse. The 3 bolts which go through the gasket also help.

27. Torque up the 3 No. Timing chest bolts.

Valve Timing
28. Hook the timing chain resting on the water pump boss and lift the chain over each camshaft.

29. Holding the timing chain tension quadrant with our favourite hook lower into the head and with some sealant round the pivot pin thread, insert the pin and tighten. The small sprocket sometimes will need replacement if worn, it is a hard working component. It is normal to stake the brass round the special fixing bolt to fully secure.

30. Place the cam sprockets on each cam in turn making sure that the correct sprocket goes on each. The exhaust cam is marked EX but it can be very faint. There is 1 tooth difference between the timing marks of the 2 sprockets.
The cams should be very close to the TDC No. 1 position with the timing marks pointing towards each other and level with the head face. Adjust as necessary.

31. Turn the engine to TDC No.1. This should be just a small turn as the piston should already be on the up stroke No. 1 cylinder. See engine step **18**

32. The chain is fitted to the exhaust sprocket first. Ensure that the chain is round the crankshaft sprocket. With the exhaust sprocket held roughly in the right timing position and keeping the chain tight, insert the sprocket into the chain and onto the cam. The timing mark should be exactly level with the top of the timing chest. Adjust the chain on the sprocket as necessary. Fit the special washer with the fixing bolt and lock washer. Please note that the sprocket bolt has a half head to keep clear of the half moon gaskets.

33. Install the inlet sprocket much the same as the exhaust. With no tension on the timing chain because the tension adjuster is not yet fitted this should be no problem. Sometimes with a new chain and sprockets it can be difficult and I find a pair of mole grips on the cam shaft just to give a tiny directional nudge if required helps. The timing marks may not exactly line up.

34. Install the chain tension adjuster, piston and spring from the outside of the timing chest with some sealant round the thread.

35. Screw the adjuster in till the chain movement between the sprockets is ½ inch total.

Note;
Tightening the chain normally has the effect of retarding the valve timing slightly. Always check the timing marks again to make sure all is correct. It is a myth that the timing marks will be exactly in line and level with the timing chest at TDC. With a skimmed head, a skimmed block, worn sprockets, and a worn chain you will be very lucky if the marks are exactly right. For a normal road going car that the marks are 'slightly' out and not in line is not important. If the marks are more than ½ tooth out always advance the inlet.

For owners that want that extra bit of horse power adjustable studs are available to allow the timing marks to exactly line up. Different after market cams will probably need adjustable sprockets as their timing will be different to the standard cams.

36. Turn the engine upside down on the stand and replace the oil suction and pressure relief pipes. Ensure that the coarse filter is properly attached and clean. The pressure relief pipe can be tapped in with a bolt inserted in the end to prevent damage. Ensure that the oil pipework and filter clear the baffles in the sump. Turn the engine over to make sure all is clear from the crankshaft.

Rear Seal Housing

There are 2 types of rear seal housings an early one with a rope seal and the later one with a lip seal. Remove the old seals. Turn the block on the stand so that the rear face is uppermost.

37(1). Lip seal Apply a coating of Wellseal to the housing and press in the seal making sure that the seal is fully in and against the stop. Grease the crankshaft where the seal runs. Apply Wellseal to the housing and block, lay the gasket on the block and bolt the housing to the block.

37(2). Rope seal The rope seal comes in 2 halves. Smear Reinzosil into the housing and push one of the seals hard into the housing. It should be about 1/16 inch proud of the face. When the silicon has cured oil the seal. Grease the crankshaft where the seal runs. Apply Wellseal to the housing and block, lay the gasket on the block and bolt the housing to the block.

38(1). (Lip seal engine) Ease some Reinzosil into the timing chest and rear oil seal housing gasket recesses. Insert the cork gaskets. Sometimes these gaskets can be a fraction wide and too tight in the recess. They can be eased by gently sanding on flat sandpaper. The fit should be good secure sliding fit, not tight so you cannot press in, not loose so it slops about.

38(2). (Rope seal engine) As for the lip seal engine but just the front timing chest. For the rear the other half of the rope seal installs in the sump. Ease some Reinzosil into the sump recess and press in the second rope seal hard. One again it should be about 1/16 inch proud. Trim as necessary. Let fully cure then oil.

39. Coat the sump and block with Wellseal. Apply the 2 sump gasket halves. The tongues on the gasket ends should go into the recesses NOT be cut off as I see with many engines. Gasket manufacturers do not make the detail bits to be 'cut off'. Lower the sump into position and install the bolts. The bolts in the block are short only ½ inch and this is important as most of the thread holes on the inlet side of the engine are blind and you do not want the bolt to bottom. The thread holes on the rear oil seal and the inner holes on the timing chest should have a longer ¾ inch bolt. All bolts should have a substantial washer to spread the load on the sump. To avoid not having the sump in exactly the correct position and then pulling the gasket about when adjusting I rest the sump on the spring loaded oil pick up and insert 2 inch long ¼ UNC bolts 2 each side of the sump just finger tight so when the sump is pushed down it is in exactly the correct position. Insert the correct bolts only removing the positional ones later. Torque up the bolts in stages. The gasket will compress and once again it is essential after a couple of hours to check tightness again.

40. Coat the head cam access recesses and half moon seals with Reinzosil. Push the half moons into the recesses. Carefully lay the cam cover gasket onto the head followed by the cam cover. Normally with a new gasket I do not use any sealant. The fixing stud holes need to be sealed or oil will seep past the nuts and make an untidy mess. Washers with a neoprene seal bonded to one side do the trick called Senloc washers or just as effective smear silicon round the stud before a washer and Nyloc nut. The gasket is a fraction unwieldy and if you just lightly pinch it with the nuts you can adjust its position in or out to make a neat job. The cam cover gasket is a thick cork item and will compress as you tighten the nuts and it is a good idea after first tightening to go round again after a few hours.

41. Install the petrol and oil pumps. It is essential that the oil pump be primed to ensure it will pick up oil immediately. The oil pump gasket should not have sealant applied to it.

42. With the engines rear face upwards on the stand install the gearbox spigot bearing. The early 4 bolt crankshaft has a sintered bronze, also called Oilite self lubricating 17mm bearing. The flat face goes towards the crankshaft, the shouldered face towards the clutch. The later 6 bolt crankshaft has a 15mm needle bearing. The O ring grease seal is fitted towards the clutch to prevent grease escaping. It is a good idea to put a bit more grease into the housing for this bearing.

Important Note.
The early gearbox with the 17mm inlet spigot and the later gearbox with the 15mm spigot cannot be interchanged without modification of the spigot bearing. If fitting a later 2000E box to a 4 bolt crankshaft engine, probably the normal mismatch, the best way to sort the spigot bearing is to machine up a steel bush in place of the Oilite bush but sized for the force fit of a needle bearing. The rarer fitting of an early box to a 6 bolt crank is best accomplished by removing the input shaft from the gearbox and machining the 17mm shaft down to 15mm and using the needle bearing.

43. Install the flywheel. The bolts should be coated with Loctite, 6 on the later Mk 2 engines, 4 on the early engines. The 4 bolt crankshaft should have a bolt locking plate but these are difficult to obtain. Loctite will ensure there is no problem in service. Make sure the faces are clean to prevent any run out. If there has been any vibration with transmission clutch operation check the surface of the flywheel and have it resurfaced if necessary. It would also be an idea to check for run out with a dial gauge.

44. Lay the clutch friction plate on the flywheel ensuring that the correct side goes towards the flywheel. The plate has a note printed on it saying 'Flywheel Side'. Install the pressure plate on the flywheel dowels and insert a clutch centralising tool. Torque up the bolts, I use a bit Loctite again here.

The engine is now ready to be mated with the gearbox and be installed in the car.

WATER PUMP REPLACEMENT IN SITU
This is the worst job to do on an Elan. There is no way to sweeten the pill, it is a nightmare. The Lotus Work Shop Manual states that the timing chest can be removed with the cylinder head in place. That is incorrect, a Lotus nonsense, the head must come off. Additionally, the sump must be dropped out. Normally on early Lotus chassis the bracing strut on the underside of the chassis prevents sump removal. Later Lotus chassis and Spyder chassis have a removable strut and this makes a water pump replacement possible with the engine in situ. Even so this job can be difficult. Some owners and even Lotus garages remove the engine for this problem as a matter of course.

On the early Lotus chassis with the welded strut it is recommended that the strut is carefully removed from the chassis with an angle grinder and a new strut of the same section, but slightly longer, is installed and bolted with 3/8 UNF bolts to the spine flanges.

It would not be the first Elan chassis I have seen with this strut cut off, then joined together with a bit a scrap metal with nuts and bolts. Not a good thing to do, not recommended, but understandable. If this has been done by a previous owner. Cut off totally and replace with a properly fabricated new strut as above. Make absolutely sure that the Chassis and new strut are well painted to prevent corrosion.

Water pump replacement with engine in the car.
1. Disconnect battery always the first job.
2. Drain and remove radiator, top heater hose and temperature sender.
3. Disconnect heater valve control cable.
4. Disconnect air box and remove carburettors
5. Drain engine oil.
6. Remove generator/alternator.
7. Remove starter motor.
8. Disconnect chassis strut (see above)
9. Remove anti roll bar.
10. Remove oil filter and sump.
11. Disconnect Exhaust at manifold, and slide off the studs.
12. Remove the cam cover.
13. Remove the spark plugs and turn the engine so that it is at TDC No1 cylinder and the valves are near enough all closed.
14. With the car in gear loosen the sprocket bolts. Release the tension on the timing chain totally and remove the cam sprockets. Let the chain drop into the chest to the water pump housing.
15. Remove the cylinder head. Undo all the timing chest bolts first and then the head bolts. Undoing the head bolts is in the exact sequence of replacement, from the centre out. Do not place the cylinder head face down on a bench, you might damage valves, support on two wood battens at either end or upside down resting on the long cam cover studs.
16. Remove the fan blade and belt pulley.
17. Remove the crank shaft drive pulley.
18. Remove all the timing chest nuts and bolts.

The timing chest is now free but unless you are very lucky it will not come off being held to a greater or lesser degree by the O ring carrier and water pump boss. Freeing the timing chest can be a tricky job. The back plate is not totally secure being held only by one bolt and pulling and levering can bend both to the detriment of those oil leaks. I have a half moon block of hardwood which goes in No.1 cylinder bore and a G clamp is fastened across this to the back plate. A mole wrench is clamped across the bottom edge. Now everything is rigid and the timing chest much easier to remove. Gripping top and bottom you can work the chest off. A pry bar inserted in the cavity can help.
Tap back the locking tabs and undo the jackshaft sprocket nuts.
Remove the sprocket.
Remove the single nut holding the timing chest back plate and remove the back plate.
Rebuild the water pump assembly as previously described.
Rebuild engine as previously described

Replacement of sump.
For the average owner mechanic working under the car the installation of the sump can be difficult. Even with a hoist I have seen garages make a mess of this job. Pushing the sump up to the block against the pressure of the oil pick up spring can cause the gaskets to move from their correct position. So as already mentioned in the Engine Rebuild Section it is better to use 2 inch long sump positioning temporary bolts.

Water Pump Replacement With The Engine Removed.
This way is probably ensures a better job but takes that bit longer. If possible if it can be done at the same time as another needy job eg. clutch replacement that is a very good idea.

OIL.
Fill the engine with new clean oil. Always replace the old oil filter. Fill the oil filter directly with oil before you screw in place. Failure to do this means your oil pump will pump oil into the filter canister before pumping round the engine. For those few seconds when your engine is running without oil pressure it is not good news.
See also '**Starting The Engine For The First Time**' below.

I have always used Classic 20-50 in my Elans. However during a recent long conversation about engine oil with Castrol their recommendation was Castrol Magnetec 10w-40 a semi synthetic oil. I have been most impressed with this product. The engine is quieter. The oil stays clean longer which means harmful products are not being carried to the bearings so wear must be less. I am looking forward to the long term results of using this oil to see exactly the full benefits.

STARTING THE ENGINE FOR THE FIRST TIME
On the twin cam engine, due to its high performance design, the valve gear and pistons want to use the same space, although hopefully at different times. It is a good idea to prevent any valve damage in the event of incorrect assembly to have turned the engine over by hand using the crankshaft pulley nut and fully check that all the timing marks are all correct. This is mentioned in the assembly text but is drawn to your attention again.

It is a good idea instead of starting the car from the cockpit to control things from the engine bay. I have made up a switch system with crocodile clips that clips to the fuse box and the starter solenoid. I can control things from the engine bay and also see what is happening at the same time. Disconnect the ignition feed to the coil whilst you are checking or the coil can get very hot without the engine running which can lead to failure.

It is another good idea, even essential, before starting the engine for the first time, after a full or partial rebuild, to pump up oil pressure initially. A bonus from this is that the carburettors are primed with petrol at the same time. Remove the spark plugs and spin the engine over on the starter until the oil pressure begins to rise. It should not take too long as the engine will spin relatively quickly without any compression to overcome, and of course, you did prime the oil pump and fill the filter with oil as you assembled it? I also have a large accurate oil pressure gauge which I connect up. The Smith's dash instrument can be a little suspect.........

Keep an eye on the petrol pump glass bowl as the engine spins over. This bowl should very quickly fill with petrol. If it does not, there is a problem. Before taking the pump off make sure the glass bowl itself is tight, the pump petrol inlet connection is secure, and the connection to the petrol tank has the clips tightened round the neoprene sleeve. If either let in air the pump will not pick up petrol but will pump air instead. You would not be the first to think that this is a pump problem when really it is a connection problem.

If your rebuild has taken some time do not use the old petrol remaining in the fuel tank. Petrol deteriorates over time and engines do not like this. The engine will cough and bang and splutter and make any attempt at checking and tuning impossible. Remember to connect the coil again and it is an idea just to check that you do have a spark at the plugs.

Install new spark plugs and start the engine. If the engine kicks back on the starter the static ignition timing is a little too far advanced. Ease a little on the distributor turning a few degrees anti-clockwise. Try again and there should not be a problem, this engine starts very readily no matter what the conditions. Check that you have full oil pressure, 40psi. Adjust the tick over to about 1500 RPM and let the engine warm up. Check for oil leaks especially the oil filter canister and petrol leaks around the pump and carburettor connections. Check for any coolant water leaks. Check the ignition timing with a timing light. This is where the dash of white paint you put on the crankshaft pulley on the TDC notch comes in useful. If the carburettors were in tune before the strip down they will be in tune fuel mixture wise now. The balance will have slipped and this should be the first thing to check. If any doubt refer to Section L, Fuel System. Adjust tick over.

CHAIN TENSION
There are 2 stages to adjusting chain tension. The chain should provisionally be tightened to the ½ inch movement stated by Lotus. When the engine runs if the chain is still rattling usually a SLIGHT ease inwards of the tension screw is enough. If the chain is whining rather than rattling it is too tight. You can shear off the drive sprockets from the cams which makes a nasty mess. A new chain will stretch when first fitted so please recheck the tension after 500 miles.

RUNNING IN.
Lotus quote 3000RPM as maximum engine speed during the run in period. Recently I had an interesting conversation with a very experienced Lotus engine designer who said that this figure came from the age when engine builders had to cut their own bearings which then did indeed need bedding in. These days with very accurate formed bearing shells these do not need bedding in, only the piston rings. So using a proper running in oil, drive normally when the engine is at working temperature, just keep the acceleration down. No starting from rest at 6000 revs!! Drain the oil out when hot at 500 miles, change the oil filter and fill with your selected oil. It is also a good idea to check tappet clearances at this mileage and generally check things over.

ENGINE PHOTOGRAPHS IN ASSEMBLY STAGE ORDER

JACKSHAFT BEARING DRIFT (Below)
The main diameter is a slide fit in the crankshaft bearing housing. The machined end holds the bearing shell. The drift is long enough to protrude from the block when inserting the centre bearing for easy access with a mallet. Makes the job easy. The drift will also remove old shells with no problem

LOTUS / FORD JACKSHAFT BEARING INSTALLATION / REMOVAL TOOL (Below)
Called a camshaft bearing installation tool mistakenly in the Lotus workshop manual and probably copied from the Ford manual where it was a 'camshaft tool.' This is slightly more complicated than the simple drift above. It can be simplified with a screw thread to hold the dolly in position and a nut in place of the handle. A ratchet and socket is almost as good. A stout flat plate will suffice as an anchor plate. The 1/2 inch diameter shaft has to be long enough to reach the centre bearing. Threads should be 1/2 inch UNF.

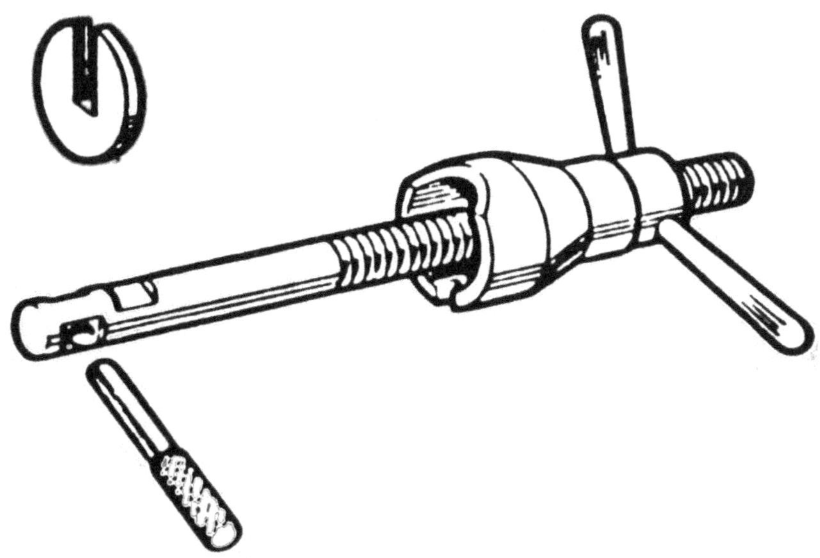

BEARING DOLLY
The smaller diameter should be a slide fit in the bearing shell and the length of the outer bearings. The large diameter should be a slide fit in the block bearing housing.

JACKSHAFT BEARING SHELLS (above)

The 3 jackshaft bearings.
The right hand side with **2 oil holes** the front bearing.
The centre with the **1 central oil hole** is the middle bearing.
The left hand side with the **offset oil hole** is the rear bearing.
It is vital that the bearing oil holes line up with those in the crankcase.

CRANKCASE FRONT JACKSHAFT BEARING HOLE

You can see the 2 oil passages. The top one goes up to the head for cam lubrication and the bottom one is the oil supply from the gallery.

You can see the 2 white paint marks I have put on to ensure the bearing shell oil holes line up.

JACKSHAFT

The Jackshaft has been slid into the block after coating the bearings with Graphogen and secured with the thrust plate which also controls the end float. Here the lock tabs on the thrust plate bolts are being carefully dressed round the bolt head. Bend up to the bolt initially with pliers.

CRANKSHAFT INSTALLATION

Install the bearing shells in the crankcase and coat with Graphogen.

Carefully lay the crankshaft into position and install the bearing caps. Check the end float.

Torque up the cap bolts to the correct 55-60 lb. ft. in 3 stages. Check for ease of rotation

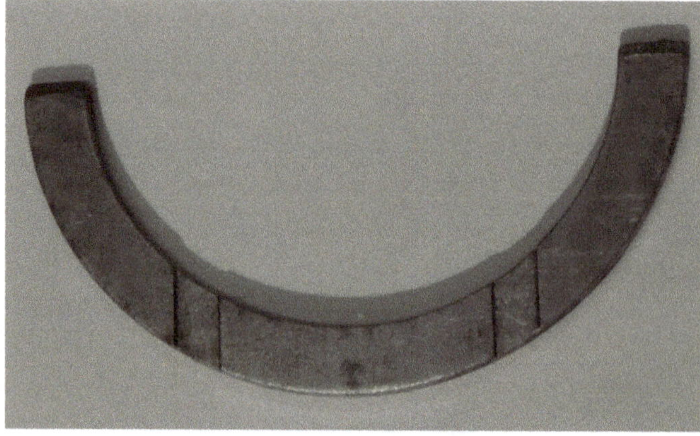

SETTING UP CRANKSHAFT END FLOAT.

The crankshaft end float should be between 4 and 8 thou. of an inch. The float is controlled by thrust washers as shown left. The washers are available in standard thickness of 93 thou. And up to plus 15 thou. in 2.5 thou. steps. It is important that the washer is correctly positioned with the oil ways facing the rotating crank.

INSTALLING A THRUST WASHER.

With the central bearing cap not yet installed the thrust washer can be fitted.

There are 2 thrust washers fitted, one each side of the central main bearing.

The washer fits in a groove in the crankcase and as it is pushed in it revolves round the crank.
See the oil ways facing correctly towards the revolving crank.

MEASURING END FLOAT

Pull the crankshaft hard in one direction and measure the float with feeler gauges.

If required install thicker washers to reduce the float. Although supplied in matched sets it is possible to fit different sizes to obtain the correct float. This engine had one standard washer and a larger one of plus 5 thou.

You can just see the end of the thrust washer next to the feeler gauge blade.

The installation of the central bearing cap holds the washers in position.

PISTON

A new piston about to be assembled into an engine. Importantly please note the directional arrow indicating the front of the piston.
The valve cut outs also give an indication of which way the piston goes, the larger cut out being for the larger inlet valve.

CONNECTING ROD (below)

A correct safer 125E con. rod ready to be mated to a new piston. A new small end bearing has been pressed in by the engine machinist.

Please note that as with the piston the con. rod has to be installed the right way which is clearly indicated. I always keep the big end bearing cap attached to the rod to make sure there is no mix up. Each cap and rod are bored to the correct diameter bolted together to ensure there is no ovality.

GUDGEON OR WRIST PIN

Gripping the con.rod in the vice, please see jaw protectors, slide the pin into the piston and through the conrod. The pin will slide easily in pushed by the thumbs. Ensure that the piston and rod are the correct way round and do not forget the Graphogen.

Secure the pin with the clips usually supplied with a new piston. I prefer the circlips to the spring clips which can be difficult to fit. Both types are shown in the photograph below.

Ensure that the ring gaps are evenly spaced before clamping the piston in the ring compressor. I have shown the compressor below upside down so you can see the indents that stop the compressor trying to slide into the bore.

GUDGEON PIN AND THE 2 TYPES OF SECURING CLIPS

PISTON RING COMPRESSOR

You need a 1/4 inch socket drive to compress.

PISTON RING PLIERS

The rings should be checked for correct gaps direct in their respective bores before fitting to the piston. Always aim for the minimum gap stated. As wear takes place and the ring expands to take up this wear, the gap will increase.

Gently expand the piston ring in the pliers just the minimum needed to ease the ring over the top of the piston. The ring metal is hard to counteract wear and therefore brittle. I have broken a few in my life.

The lower oil control ring is first followed by the 2nd ring and then the top compression ring. Please make sure that the rings are the right way up. Normally the manufacturers name and ring size are etched on the top face.

FITTING PISTON RINGS

I hold the piston by the conrod very lightly in the bench vice just to steady it whilst easing the rings over the top.
You can see here that 2 rings have been already been fitted. All the books say that the gaps should be spaced equally round the piston before fitting in the block bore and that is certainly correct advice. The rings actually revolve in use and how long they stay like that is not known to me. Please note the cleanliness. Not quite Formula One but very reasonable!!

INSERTING THE PISTON

With the ring compressor tight around the piston and Graphogen lubrication applied to bore and piston AND making sure the piston is round the right way ease the piston into the crankcase bore.

The piston should slide down easily to the rings when the gentle tapping of a mallet handle will fully insert the piston.

TIGHTENING THE BIG END CAPS.

With the crank shaft at BDC guide the conrod to the journal.

Install the bearing cap having first applied Graphogen to the bearing surface.

Torque up the bolts.

Turn the crankshaft to ensure there is no binding.

TIMING CHEST BACK PLATE

The back plate fitted to the crankcase and the sprocket fitted to the jack shaft. Please note the locking washer with the tabs bent up against the bolts. You can see clearly the one bolt that holds the plate to the crankcase.

Please also note the cleanliness

WATER PUMP

No self respecting engine builder would insert the pump dry these days as Lotus suggest in their workshop manual. I apply Reinzosil round the O ring carrier. Not only does this make a sure seal but also provides a little lubrication for the O ring to slide easily into the back plate. You can also see the paper gasket that I make up, this will be a working engine.

TIMING CHEST

Wellseal applied round the edge seal face. I use a 30 thou. paper gasket here which is laid on the back plate. To keep the sealing faces true a similar thickness gasket, mentioned above, has to be made up for the pump boss which you can see fitted. This timing chest is ready to be installed into the back plate. With no gasket I would have used Reinzosil on the aluminium faces. You can just see the Reinzosil round the O ring carrier.
Note again the cleanliness.

BACK PLATE

The back plate ready for the timing chest. Wellseal has been applied to the edge seal faces ready to lay on the gaskets. Reinzosil has yet to be applied to the pump boss both on the inside and on the face. The timing chain has been fitted CLEAR of the pump boss.
The oil slinger has been fitted. Having the chest turned horizontal on the engine stand helps keep all in place

Carefully push the timing chest into the back plate and install the bolts.
See below for tightening.

WATER PUMP BOLTS (Below)

The diameter of the 3 No. Bolts round the water pump boss, 2 are 1/4 diameter UNC thread whilst the third, lower right, is 5 /16. The bolt which holds the generator belt tightness adjustment is 5/16 diameter but not shown as not installed yet. All the rest are 1/4 UNC or UNF. However,
IMPORTANTLY,   despite the difference in size all should be torqued to equal tightness of 6 lb.ft to give equal pressure over the sealing surfaces.

It is a good idea to trim the edge gaskets flush with the top (head) and bottom (sump) faces so ensure no interference.

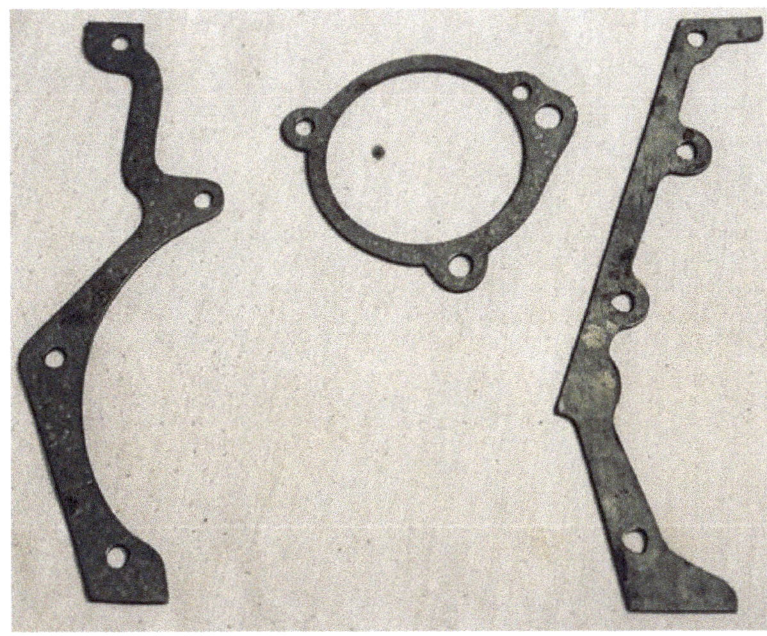

**TIMING CHEST
GASKETS.**

(left) A pattern for the gaskets made from sheet metal and the resultant gaskets in the second picture below. Draw round the pattern, cut out with a 'surgeons' knife and cut the holes with a hole cutter.

This does keep the oil seepage under control and I have used these gaskets for something like 250,000 miles without a problem. If you 'dome' silicone and let it cure the gasket is just as thick but not as good.

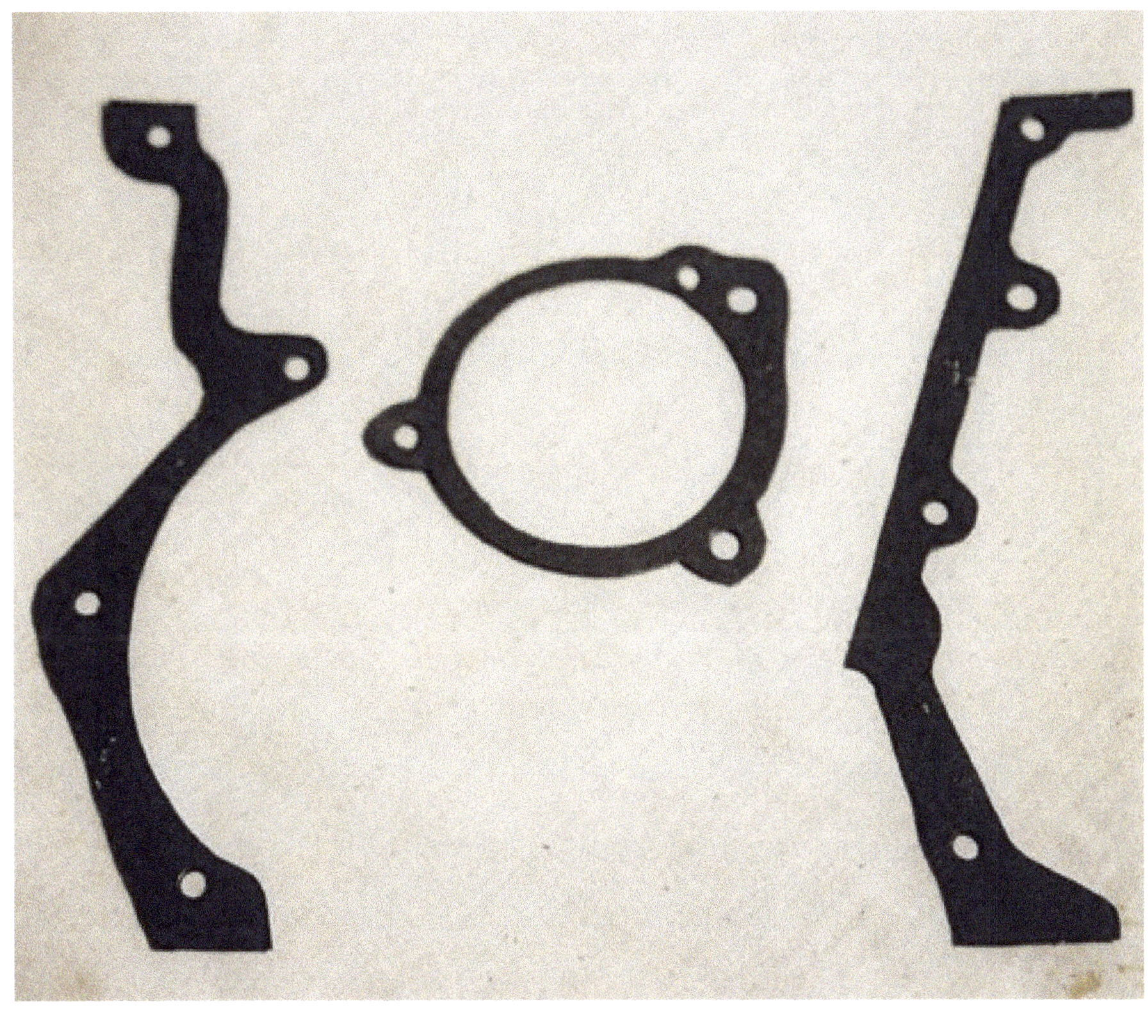

ENGINE MAIN GASKETS

Left The paper timing chest back plate to crankcase Gasket.

Below The timing chest to head gasket.

Above The copper/aluminium head gasket

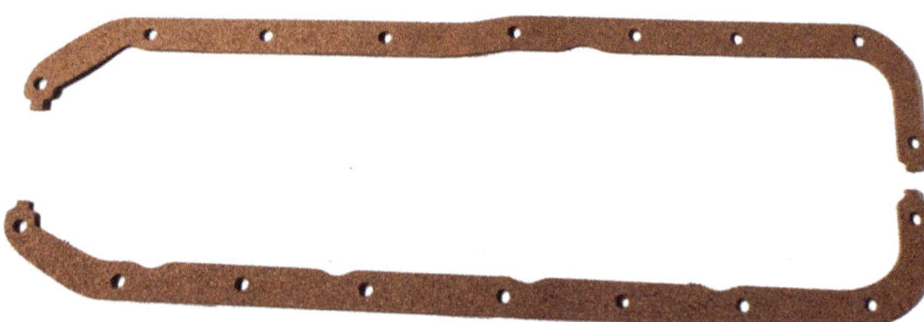

Left. The Mk. 2 engine cork sump gasket. The Mk. 1 is very similar just a different size at the rear.

Left. The very flimsy cam cover gasket. Please take note of suggestions in text regarding fixing.

Below. The head half moons

CYLINDER HEAD INSTALLATION

There are so many things to think about at the cylinder head stage so let's do it the easy way. Turn the engine so that the pistons are about mid position, on the upstroke cylinder No. 1. Using Wellseal fix the cork gasket to the top of the timing chest. Using a good amount of silicone attach the rubber breather pipe to the block. Screw locating studs to the crankcase also holding the head gasket. Smear corresponding Wellseal and silicone sealant on the head and ease over the studs and lower onto the spacers. I use 5/16 diameter rod for spacers. Put all the head and timing chest bolts in just the first few threads. Making sure that the breather pipe lines up, lift the head a fraction, one end at a time and pull out the spacers. The head will now be down onto the block and in the correct position. Remove the locating studs and replace with the proper head bolts. Ease up the chest head bolts first just so that they pinch the cork gasket. Make sure the gasket stays in the correct position. Torque up the head bolts to 65lbs. ft. in the order as stated in the Lotus manual, do it in at least three stages. Torque up the 3 No. timing chest bolts to 12lbs. ft. Sometimes I say it is not skill it is technique.

CYLINDER HEAD INSTALLATION 2

1. (Above) Chest bolts inserted whilst spacers still in position. It is a good idea to coat all head threads with Copperslip.

Note. The timing chain tensioner bracket pivot pin is missing. It is easier to install the head without the bracket and sprocket dangling down and they are easy enough to install with the head bolted to the crankcase.

Note. The timing chain tensioner adjusting screw right out ready for the camshaft sprockets and valve timing.

2. (Left) The rubber breather pipe securely located in the head with a good sealing of Silicone. If you forget to install the breather pipe with the head, there is no way that it can be installed retrospectively. Take the head off and start again, with new gaskets of course. It is expensive to make mistakes like that.

VALVE TIMING

With the head installed the next job is valve timing. With a bent bit of welding rod or long nosed pliers, hook the chain tensioner and lower into the head and fasten with the pivot. A good bit of Reinzosil round the thread of the pivot will stop all oil leaking out. At this stage do not install the chain tension adjusting piston. This is to ensure that there is no tension on the chain

LIFTING THE TIMNG CHAIN

(Below) With the same bit of bent welding rod hook the chain resting on the water pump boss and lift the chain over the cams. The cams should be near enough at TDC but now the cams should be adjusted to exact correct TDC position by putting the sprockets on each cam and adjusting the timing marks. A Mole wrench on the cam just to nudge it makes this easier.

ENGINE TIMING

With the cams adjusted turn the engine with a socket on the pulley bolt so that the timing notch on the pulley exactly lines up with the TDC mark on the timing chest. Note: please see the white paint on the pulley timing notch. This makes it far easier to see when setting the dynamic ignition timing with the strobe light.

Both the engine and head are now at the correct position at TDC No. 1 for the chain to be installed over the sprockets.

EXHAUST SPROCKET
The chain is installed over the exhaust sprocket first. Keeping the chain tight from the crankshaft sprocket insert the exhaust sprocket into the chain so that the timing mark is level with the head top face

INLET SPROCKET
Install the inlet sprocket. With both sprockets connected the chain starts to tighten and sometimes you have to give the cam a bit of a nudge with the Mole wrench to get the cam and sprocket to line up. With worn sprockets, maybe a worn chain and definitely the head skimmed after some 50 years, the timing marks are not going to line up exactly.

CHAIN TENSION
Install the chain tension adjusting mechanism from the outside and tighten the chain to the recommended 1/2 inch movement. This adjustment will retard the timing so please check the timing marks after.

ALWAYS CHECK (Below)
So with the chain adjusted the timing marks are not far out but always check.

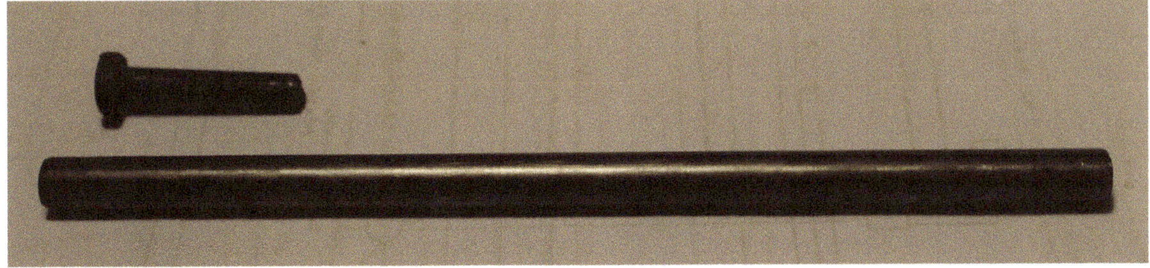

OIL PUMP PIPEWORK

The suction (large) and return oil (small) pipes on a Mark 2 engine tap into the crankcase with a little Wellseal applied round them.

The pipes are a gentle force fit. They really do tap in. However to prevent damage to the pipes it is normal to use a bolt filed down to a nice slide fit into the tube. **See above**

Left. This suction pipe was damaged in removing from its original block and you can see the new braze on the washer base for the spring. This pipe has to be rotated to the correct position to that the filter plate traps the return pipe. The filter has also got to be positioned to avoid the anti surge baffles in the sump. Rotate the engine to make absolutely sure that all is clear from moving parts.

The security tabs for the coarse filter are a bit fragile and one broke off whilst straightening. A repair job for me before the sump goes on.

The early Mark 1 engine has a slightly different design in that the suction pipe is held in the crankcase with a back nut. The filter is much larger diameter but the principle of positioning to miss the sump baffles are the same. The baffles in these engines are different anyway to the later Mark 2.

COARSE OIL FILTER
The coarse filter fitted with the tabs bent over the bottom edge to safely secure it. I have even repaired the RHS tab. Please note the spring which I will refer to when fitting the sump.

MARK 2 ENGINE SUMP
Photograph showing the anti surge baffle plate. It is this baffle plate that you have to be careful to avoid when fitting the oil pump pipes.

SUMP GASKETS
Above. The 2 cork gaskets which seal each end of the engine
 Left The rear lip seal housing showing the recess where the large cork half round gasket inserts. The smaller cork gasket inserts in a very similar housing contained in the timing chest sealing the front of the engine.

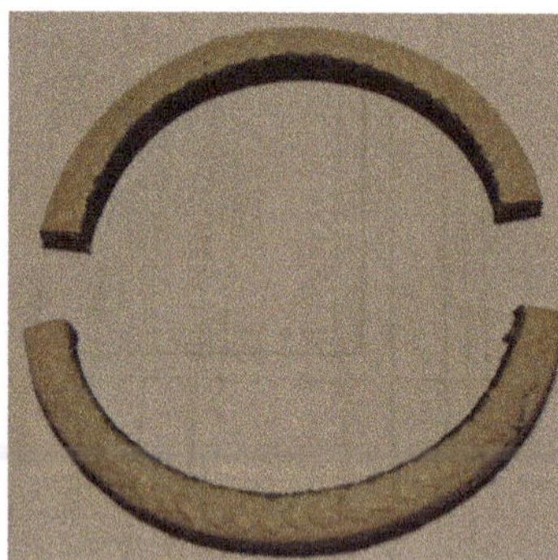

MARK 1 ENGINE
Left. The early engines had a rope seal round the crank-shaft which was tricky to fit but described in the text. Ford must have had some trouble and changed it to a lip seal type on the later Mark 2, 6 bolt crankshaft engines.

INSTALLING THE SUMP

Apply Wellseal to the sump including the half round sections for the end seals

Apply Wellseal to the crankcase including the recesses for the end seals.

Lay the sump gaskets on the block. Make sure that all the holes for the sump bolts line up. You will have noted that the 2 gasket are slightly different. There is a small tab at each end and it is impostant that this fits into the recess housings at each end.

Push the half round cork gaskets into the recess at each end. They finish on top of the sump gasket.

PLEASE NOTE When the sump is placed in position the spring on the suction oil pipe keeps the sump from sitting on the block. I use 4 long bolts to hold the sump in the exact correct position when I push it to the block. You do not want the gaskets to be pushed round. This is even more of an assistance if installing a sump to an engine upside down on a ramp.

SUMP BOLTS

There are 18 1/4 inch UNC bolts holding the sump. 14 are 1/2 inch long and 4 are 3/4 inch long next to the end seals. All have a large locking washer ....large to spread the load.

INSTALLING THE CAM COVER

Above left. Apply Reinzosil to the half moon cut outs.

Above Right. After applying Reinzosil to the rubber half moons fit into the cut outs.

Left. Lay on the cork gasket

KEEPING THE OIL IN

Oil will weep out under the 8 fixing nuts if you are not careful. You can use Senloc washers, a washer with a sealing rubber disc attached to it. It is cheaper and better in my opinion to use Reinzosil applied round the stud, then a standard washer before the 1/4 UNF Nyloc nut.

The gasket is large and 'floppy' and sometimes for neatness you have to nudge bits of it into the correct place the weight of the cam cover will normally hold it but if not a slight pinch up of the nuts will do the trick.

Tighten the nuts evenly, going side to side and back to front. It is a thick gasket and will compress somewhat. Leave it for a couple of hours and then go back and retighten.

Please do not forget to clean off the Reinzosil before curing.

FITTING THE FLYWHEEL

Make absolutely sure the flywheel bolt boss is clean. Dirt between the crankshaft and flywheel will encourage runout giving clutch take up problems. Make sure the special bolts are clean and dry. Oily bolts do not encourage the Loctite to stick.

FLYWHEEL BOLT

The flywheel bolt is a high tensile special bolt with a shoulder to make absolute sure there is no movement of the flywheel. The thread is 3/8 UNF and they should be torqued up to 45-50 lb.ft.
Loctite should be applied.

4 BOLT CRANKSHAFT

This is the bolt locking plate from an early 4 bolt crankshaft. A little bit tatty where the tabs are bent over the bolt but this plate was the only one I had and as stated in the text they are no longer available. A spring locking washer and Loctite is a good answer.
This plate could easily be made from a piece of flat steel plate. The 4 bolts should be hexagon head to ensure good locking from the bent up tabs.

LOCKING THE FLYWHEEL.

When tightening or removing flywheel bolts the flywheel has to be locked or the engine will turn. Several methods can be used but I have stuck with this one over the years which is totally hands free. The lock mechanism uses one of the bell housing bolt holes.

The holding bolt should have a nut on it and I was obviously being lazy here. See photograph below.

FLYWHEEL LOCKING TOOL.

Made up by myself in a few spare moments, I have used this flywheel locking tool for a few years. The spacer to allow the teeth to engage exactly with the flywheel is a spare left hand engine support spacer cunningly reused.

FITTING THE CLUTCH

Put the friction plate on the flywheel ensuring that it is the correct way round to the flywheel. It is printed on the plate 'FLYWHEEL SIDE'. Fit the pressure plate over the flywheel studs.
Centralise the friction plate with a clutch tool. Accuracy with this job will ensure an easy fit of the gearbox to the engine. Ease the bolts up equally keeping a wary eye on the centralisation.
For more information on the clutch see the CLUTCH SECTION Q.

ANCILLARY ITEMS

OIL PUMP

Installation of the oil pump is no problem. A simple gasket and 3 bolts. Most professional builders do not use sealant on the gasket to ensure that no squeezed out residue gets into the pump.

I am not a fanatic about always using a new pump on a rebuild. The oil pressure gauge is there to tell you if there is a problem.

I am a fanatic about changing oil **regularly**, with a new filter.

PETROL PUMP

Installation of the petrol pump is easy whilst on the engine stand. In the car the chassis gets in the way.

The pump depending on the position of the cam on the jack shaft has to be pushed downwards to line up with the fixing holes.

Sometimes I say it is easier to turn the engine so that the cam position is in the non operational position. The cam is easy to feel with a finger in the pump crankcase hole.

In this position the pump hardly has to be pushed downwards at all.

The 1/4 inch thick gasket is there to prevent transfer of engine heat to the pump and therefore to the petrol.

It is a good idea to check that the pump is working. They do pump well if you operate the lever by hand with the suction in a bowl of water.

A FEW MORE ITEMS TO HELP.

REPLACING TIMING CHAIN IN SITU.
The original timing chain was a continuous chain which could only be replaced with a full timing chest strip down as with replacing a water pump. Split chains joined together with a joining link and spring clip have been available for some time. Although frowned upon by some people I have had no problem with them and my own Elan has used them. They make changing a chain in service very easy. With a full rebuild it is better to use a continuous chain and only use a split link one if needing to replace in service.

Turn the engine till in the cams are pointing towards the plug well No. 4 cylinder, which is TDC No. 1
Remove the sprocket bolts.
Unscrew the timing chain tensioner fully.
Remove the sprockets.
The old chain has to be cut with an angle grinder. Tie a piece of string each side of the proposed cut and secure to prevent you loosing the chain. Cover the engine with a cloth to prevent metal swarf entering the engine.
Attach new chain to the old one and using the old chain pull the new chain round the engine. Tie the string to the new chain. Join the new chain together. The **U** bend in the spring clip MUST point in the direction of engine rotation.
Re-time engine

It is an easy job tensioning the timing chain, the adjuster is screwed in until the correct chain tension is obtained. For major work always have the cam cover still removed for tensioning. For slight trimming of chain tension there is no need to remove the cam cover. Use a stub screw driver for the adjuster screw. A point to watch is tightening the Nyloc locking nut at the end. They tend to turn the adjusting screw, tensioning the chain even more. The stub screw driver held in the slotted adjustment screw end as you tighten the lock nut with the ½ inch spanner is the way to do it.

The sprocket on the timing chain tension adjustment also wears and can be replaced relatively cheaply.

PISTONS (*).
Although the original style pistons are very difficult to get, suitable replacements are available without a problem in standard diameter, plus 20thou.(0.5mm), plus 30thou.(0.75mm), plus 40thou.(1.0mm) and plus 60thou.(1.5mm) Liners are also available and in theory you should be able to keep your twin cam going until the petrol dries up. I had my own car fitted with liners a few years ago and re-bored back to standard pistons.

PETROL PUMP (*)
The petrol pump is a standard Ford component. If building an engine out of the car there is no problem fitting this unit. There can be a problem with replacement with the engine in the car with a Lotus chassis. The flange of the chassis gets in the way. However, all is not lost. With a flat board under the sump lift the engine with a trolley jack to just take the weight off the engine mounting brackets. Remove the brackets and push the engine over to the left hand side of the car. Now you can get the petrol pump in and out. With

the chassis in the way it can be a bit awkward replacing the pump especially if the cam is upwards requiring you to hold the pump down against the spring loaded lever as you insert the fixing bolts. Turning the engine so the cam is at the bottom of its stroke does make it slightly easier.

The ¼ inch thick gasket is essential to prevent heat transfer from the engine to the pump and thus the petrol!!

The suction connection to the pump can be tricky. Use an old ½ inch AF. spanner which has been cut in half, this will be even smaller than a proper stubby spanner.

The flexible connection from the pump discharge to the carburettors is under pressure and it is essential that this connection is sound. Metal braided connections are best. If there is any fuel leak you must remember that this area is right above the coil and many Elans have come to a fiery end due to this layout. Not withstanding the petrol, glassfibre resin also burns very well. If your flexible connection is not sound, some dealers will fit new pipes on the fittings for you, which is slightly cheaper than a new component. The banjo connections need to be tight and with a full rebuild new fibre washers would be a good idea too.

It is the pump spring that controls the pressure of the petrol to the carburettors. The Lotus manual says 1.5 to 2.5 psi. High pressure pump springs are available if required for racing. If these pumps are fitted to a standard car the higher pressure petrol can force itself past the needle float valves in the carburettors and out of the float chamber vent. With the coil directly below the result is not hard to imagine.

These pumps with the passage of time are becoming increasingly difficult to obtain. A replacement is obtainable new but is not the same style as the original glass top so "originalists" may frown. Some owners are fitting electrical pumps in the boot / trunk of the car which has the advantage of pumping petrol to the carburettors without turning the engine over on the starter and flattening the battery. The disadvantage is that the all the petrol line is now pressurised and the plastic fuel line should be replaced by a metal one. It is interesting that on the original S1 Elan the fuel line was a metal tube. It would probably be better engineering to fit the electrical pump in the engine bay.

OIL PUMP

The oil pump is a standard Ford Kent pump with the pressure relief spring modified to give an oil pressure around 40psi. Maybe a bit less on worn units. Any large difference indicates a problem which should be investigated. Make sure that the oil filter is clean before you start worrying and it may also be a good idea to fit a new pump before stripping down. In the life of my car I have fitted 2 new pumps. If there is an engine bearing wear problem normally it is audible on cold start up.

The same problems of oil pump replacement in situ apply as the petrol pump and the fitting technique is exactly the same.

DISTRIBUTOR. (*)

There were several different Lucas distributors used in the Elan engine. Please see Chapter 5, Obsolete Parts. The difference between the various distributors is the centrifugal advance of the timing. Generally you can say at higher engine revs. The fuel mixture has less time to burn so the firing of the engine should be more advanced. The engine compression also comes into the equation but Lotus have sorted all that out for you. Please ensure that the distributor you have matches your engine. See the Lotus Manual Technical Data.

The original ignition system was obviously points and a condenser. Unfortunately these are only available today manufactured in third world countries and their quality is suspect. I have no hesitation therefore in recommending the use of an electronic system most of which are extremely good. I recommend an Aldon unit which fits totally in the existing distributor cap and only an Elan Sherlock Holmes would be able to spot the modification. Apart from better firing the further advantage of electronic systems is you do not have to remove the awkward positioned distributor every 6000 miles to replace/adjust the points and slight wear on the bearings does not affect the firing.

The adjustment of the distributor points can be checked very simply with a dwell meter and they should be 60 +/- 3 degrees. The distributor points cannot be adjusted or replaced in the car the distributor has to be removed. This means losing your ignition timing. Not all owners have a dwell meter or a strobe to adjust the timing. These instruments are available fairly cheaply these days and are essential if you maintain your own car.

Slacken the 7/16 pinch bolt on the clamp and withdraw the distributor to the bench. Install new points and set the gap to 15 thou. I always fit new points. At £3/4 a set it is just not worth playing with old points.
Check that the points work with a multimeter across terminal and earth.
Turn the engine to 10 degrees BTDC No. 1 cylinder on the engine timing marks on the timing chest.
Most owners mark on the distributor body the position of No.1 cylinder HT lead. Turn the rotor arm to point to this position but allow for the arm to turn about 30 degrees clock wise as the distributor is insertd into the skew gear drive.
The distributor clamp does not need to be bolted up very tight. You can easily crack the casting if over tightened
Replace all HT. and LT. connections.
Start the engine and check ignition timing is 10degrees BTDC. Check the dwell angle. If the dwell angle is outside the stated limit remove the distributor and re-adjust.
Remember the larger the gap the smaller the dwell angle.

The Standard HT. ignition cables are copper cored. The coil is a standard component, and normally there is no ballast resistor in the ignition system.

TIMING MARKS AND BELT DRIVE PULLEY.
The ignition timing marks are on the front of the timing chest just on the edge of the fan belt drive pulley. Early timing chests had three marks, TDC., 10 and 20 degrees BTDC. Later units had a fourth mark for 30 degrees BTDC.

Getting a timing strobe light to shine on the marks can be difficult. There is not a lot of room between the engine and the radiator and the radiator fan is also revolving. The marks are much easier to see if you apply a touch of white paint on the marks and also the notch on the pulley.

PLUG THREADS.
Plug threads can be re-threaded with a Recoil unit in situ if great care is taken.

1. Turn the engine to midstoke on the faulty plug threaded cylinder.
2. Liberally smear the tap with clean grease. The tap is a two stage tool, the first stage being a standard plug tap and the second a larger tap for the Recoil unit.
3. Begin to re-tap the plug hole. Remove the tap every one turn, clean off the swarf and re-grease.
4. Continue until the new larger thread has been cut.
5. Turn the engine to TDC on the faulty plug hole and with a torch check if any swarf has dropped onto the piston. Only once has this happened to me and it can be removed with a dowel with grease on it.
6. Smear high temperature RTV silicone on the coil and screw into the plug hole.
7. Using long nosed pliers so that the tang does not drop into the cylinder, break off the tang.

Using plenty of grease is the secret of this job and it does save a wearying strip down of the head. A further description of this job is supplied with the Re-coil plug tool. The job takes just under an hour to do and it does not take many stripped threads to pay for the tool about £65.

STUD THREADS.
Manifold stud threads especially on the exhaust have a habit of stripping and the 5/16 UNC. Recoil kit is needed here.

BREATHER PIPE.
Early engines had a metal breather pipe running from the head just behind No. 4 induction track alongside the engine and fixed to the gearbox. It is a good idea to remove this if on your engine and run as on later engines to the induction air box. The vacuum in the air box will slightly help reduce the crankcase pressurisation which is no bad thing and better for the environment too.

TESTING THE ENGINE.
If you are buying an Elan in running condition and require an assessment of the state of the engine the following will help you form an opinion.

Whilst the engine is cold have a good look round it for oil leaks. Most leak to a greater or lesser extent. The early units with the rope crankshaft main seal probably more than the later lip seal engines. Check the water pump by testing a fan blade for movement. Check how much the timing chain tensioner is wound in. If it cannot be wound in anymore the timing chain is at the end of its life.

On starting there should not be any undue clattering of the starter gear which would indicate either Bendix spring and gear on the starter motor or worse the flywheel starter ring worn. Some of these engines have done a few miles. It could be that the starter motor is just loose, the top bolt is difficult to get to and this is quite common.

If the engine knocks a little when cold and this disappears when hot that is a sure sign of wear especially in the bearings.

TOP DEAD CENTRE.

Cam lobe position No. 4 Cylinder (above)
If you require TDC. No. 1 power stroke for static ignition timing, the best way to check that you are not on TDC exhaust stroke, is to look at the cam lobes on No. 4 cylinder through the oil filler hole. Both lobes should point together as with the sprocket timing marks.

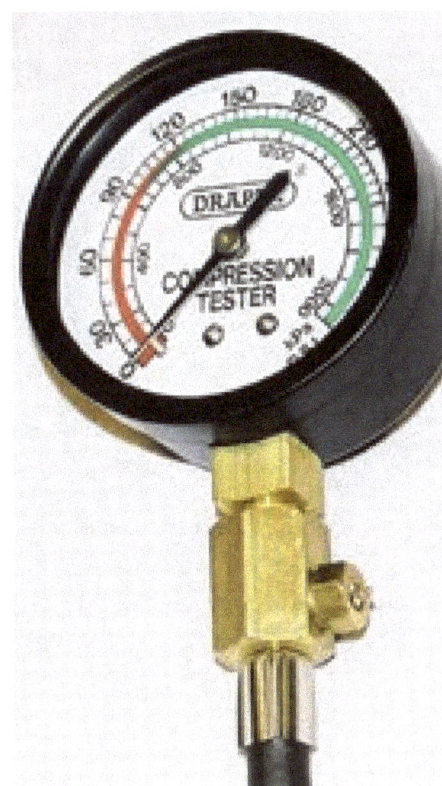

COMPRESSION ENGINE TEST

The throttle MUST be wedged wide open for a compression test. If you do not let the air into a cylinder the compression will be incorrectly low and lead to incorrect diagnosis. Remove all plugs so that the engine will spin over quickly without a problem.

On any problem cylinder put a small amount of EP 90 oil into the plug hole and test again. The thick oil will seal the piston and if the compression on retest rises significantly, piston ring problem is suspect rather than valves.

Cylinder pressures should be within 20 psi of each other;
Standard 9.5 compression ratio greater than 160 psi.
Sprint 10.5 compression ratio greater than 170 psi.

The engine should tick over quite happily when hot at 850 / 900 rpm. Normally if they will not the carburettors have to be the first port of call, especially bearing in mind the age of some of them today. Yes check the tuning, the balancing first but carburettor restoration could be rearing its head.

On a test drive use fierce acceleration after running with the throttle closed on overrun and look in the rear view mirror for blue smoke. This is oil burning and is indicative of wear somewhere letting oil into the combustion chamber. A plug inspection gives a good indication, wet carbon is a sign of burning oil. Black dry sooty plugs is bad combustion. Do not confuse them.

A compression test will help you pinpoint the problem. A compression test is done with the throttles full open. If you do not allow air into the cylinder the compression will be low. All compression test figures given by manufacturers are based on open throttles and a warm engine. The Elan should give pressures of ;

Standard 9.5 compression ratio engine............   In excess of 160 psi.
Sprint   10.5 compression ratio engine.............   In excess of 170 psi.
All cylinders should be within 20 psi.

If low pressures are obtained put some thick EP90 differential oil in the combustion chamber and immediately test again. If the pressure rises this indicates worn piston rings and that a full rebuild is required. If the pressure readings stay the same then valve problems should be suspected.

A zero reading normally indicates a piston hole. It could be a burnt valve but normally when a valve goes it is a slow drop off in performance as more wears takes place so you do get some compression. However a hole in a piston is zero.

ENGINE PAINT.
When the engine rebuild has been completed you will want to paint the unit 'Lotus Grey'. There have been many stories about the shade of this paint about how young lads were sent from the factory with petty cash to buy what they could. However, despite this, the engine colour remained remarkably consistent over the years.

The grey has been described variously as Massey Ferguson tractor grey, battleship grey or even smoke grey.

QED stock tins of paint for re-sale.

I use Dark Camouflage Grey, single pack enamel BS381C 62

BALANCING.
When rebuilding a new engine there always seems to be discussions about the necessity of engine balancing. For normal roadwork it is not necessary. If you buy 4 pistons from the same manufacturer they will be within the desired tolerance of weight. The same with connecting rods. The clutch will be balanced. If you want a 'super' engine or wish to go racing then balancing will be required.

MODIFICATIONS.

Camshafts.
People want more power to use the Elan's roadholding and the easiest way to begin is to install different profile camshafts. The standard Elan is very mildly tuned by today's standards and Lotus's own later camshafts will certainly provide you with more BHP. 2/3 of the increase in power of the Sprint Big Valve engine comes from the camshaft, the rest coming from valving, higher compression and carburation. There are many engine shops that will supply you with camshafts giving greater power. Please remember that as the power output increases the engine will become less tractable for normal road use.

A final warning note on modified / new cams. Some greater power cams require different valve springs and collars or else the valve spring will bind. This tends to make a nasty and expensive mess of the engine. Please check clearances before trying to start the engine.

Valves.
Larger valves and seats exactly as the Big Valve engine can be retro fitted. If you have hardened valve seats and new bronze valve guides, you can now use lead free petrol.

Water Pump.
There is a modified timing chest available that allows you to remove and replace the water pump without all the drama of a total strip down, just 3 No. bolts the manufacturers say. Expensive at around £600 there are not many on Elans. They did have a reputation in the early days of the unit for bad sealing but the basic twin cam had that problem itself with the O rings. No self respecting builder would rebuild an engine today without a good wipe of silicone round the sealing O rings.

Backbone chassis Immensely strong and torsionally stiff welded steel backbone carrying all mechanical components.
All independent suspension Front suspension by unequal length wishbones, combined coil spring/damper units. Independent rear suspension by Chapman strut system incorporating wide based lower wishbone and combined coil spring/damper unit.
Disc brakes Hydraulically operated calipers on 9½ inch. diameter discs on front wheels, 10 inch diameter discs on rear wheels.
Twin cam engine Light alloy twin overhead camshaft cylinder head 1558 c.c. Compression ratio 9.5:1, B.H.P. 105 @ 5,500 r.p.m. Torque 108 lb/ft @ 4,000 r.p.m. 5 bearing crankshaft. Two twin choke 40 DCOE2 Weber carburettors.
Close ratio gearbox Gearbox ratios:—1st 2.50:1, 2nd 1.64:1, 3rd 1.23:1, 4th 1.00:1, rev. 2.81:1 (3.9:1 final drive ratio).
Performance 0-40 - 4.0 secs. 0-60 - 7.4 secs. 0-80 - 13.8 secs. max. speed 115 m.p.h.
Price Basic price £1148 inc. heater and close ratio gearbox, plus purchase tax £239.14.7. Hard top an optional extra at £68.

The sleek and aggressive glass fibre reinforced plastic coachwork which further enhances the advanced mechanical specification, is included at no extra charge.

LOTUS CARS LIMITED · DELAMARE ROAD · CHESHUNT · HERTS distributing through : Ashmore Brothers Ltd., West Bromwich · Automobile Centre Ltd.. Leeds   Moto Baldet Ltd., Weston Favell, Northants   Chequered Flag (Grand Touring Cars) Ltd., Edgware   Coombs and Son Ltd., Guildford · Dicksons Motors (Perth) Ltd., Perth